A WORLD BANK COUNTRY STUDY

Angola

Oil, Broad-based Growth, and Equity

THE WORLD BANK
Washington, D.C.

Copyright © 2007
The International Bank for Reconstruction and Development/The World Bank
1818 H Street, N.W.
Washington, D.C. 20433, U.S.A.
All rights reserved
Manufactured in the United States of America
First Printing: May 2007

 printed on recycled paper

1 2 3 4 5 10 09 08 07

World Bank Country Studies are among the many reports originally prepared for internal use as part of the continuing analysis by the Bank of the economic and related conditions of its developing member countries and to facilitate its dialogs with the governments. Some of the reports are published in this series with the least possible delay for the use of governments, and the academic, business, financial, and development communities. The manuscript of this paper therefore has not been prepared in accordance with the procedures appropriate to formally-edited texts. Some sources cited in this paper may be informal documents that are not readily available.

The findings, interpretations, and conclusions expressed herein are those of the author(s) and do not necessarily reflect the views of the International Bank for Reconstruction and Development/The World Bank and its affiliated organizations, or those of the Executive Directors of The World Bank or the governments they represent.

The World Bank does not guarantee the accuracy of the data included in this work. The boundaries, colors, denominations, and other information shown on any map in this work do not imply any judgment on the part of The World Bank of the legal status of any territory or the endorsement or acceptance of such boundaries.

The material in this publication is copyrighted. Copying and/or transmitting portions or all of this work without permission may be a violation of applicable law. The International Bank for Reconstruction and Development/The World Bank encourages dissemination of its work and will normally grant permission promptly to reproduce portions of the work.

For permission to photocopy or reprint any part of this work, please send a request with complete information to the Copyright Clearance Center, Inc., 222 Rosewood Drive, Danvers, MA 01923, USA, Tel: 978-750-8400, Fax: 978-750-4470, www.copyright.com.

All other queries on rights and licenses, including subsidiary rights, should be addressed to the Office of the Publisher, The World Bank, 1818 H Street NW, Washington, DC 20433, USA, Fax: 202-522-2422, email: pubrights@worldbank.org.

ISBN-10: 0-8213-7102-9   ISBN-13: 978-0-8213-7102-2
eISBN: 978-0-8213-7103-9
ISSN: 0253-2123   DOI: 10.1596/978-0-8213-7102-2

Library of Congress Cataloging-in-Publication Data has been requested.

Contents

Preface vii

Acronyms and Abbreviations ix

Prefácio xi

Sumário Executivo xiii

Executive Summary xxxv

Introduction 1

1. **Country Background: Socio-Economic Realities Before and After Independence** 5
   Socio-Geographic Characteristics 5
   Living Standards Indicators 7
   The Transition to Independence 11
   Policy Choices and Structural Changes 14

2. **Macroeconomic Performance in a Time of Transitions** 19
   The Transition to a Market Economy 19
   More Revenues and Less Inflation 22
   The Tensions of Transition 32

3. **Oil Wealth: Policy Options to Manage the Windfall** 39
   The Characteristics of the Petroleum Sector 39
   The Legal Framework Governing the Sector 43
   Governance, Transparency and Institutional Capacity 47
   Oil Wealth: How Much and for How Long? 53
   Intergenerational Considerations 59
   Policy Options To Manage the Windfall 60

4. **The Diamond Sector: A Potential Underexploited** 67
   The Characteristics of the Diamond Sector 67
   Governance: Opaque and Unstable Legislation 69
   The Business Environment: Not Competitive Enough 71
   A Three-Pronged Strategy to Unleash the Potential
   of The Sector 73

5. **Private Sector Development and the Business Environment** 81
   The Private Sector in Angola 81
   Trade Patterns and Regional Integration..................................... 88
   Recent Actions to Improve the Business Environment 93
   The Way Forward.. 96

6. **Removing Obstacles to Agriculture and Rural Development**................. 103
   A Sector Facing Daunting Challenges 103
   Removing Obstacles to Output Growth 105
   Stimulating Competitiveness through Better Incentives 113
   Annex I... 116
   Annex II.. 117

7. **Supporting Livelihoods and Improving Service Delivery**................... 123
   Supporting Livelihood Strategies.. 123
   Strengthening Existing Social Programs 127
   Reaching the Poor with Social Services 130
   Using Fiscal Savings to Improve Service Delivery.......................... 135

Statistical Appendix 139

References 153

Map of Angola ....

Endpiece..

List of Tables

E.1 A Scorecard to Assess Governance and Transparency in the Oil Sector xlvi
E.2 Summary of Diagnostics and Recommendations li
1.1 Basic Poverty and Social Indicators 8
1.2 Composition of GDP by Sector, 1966–2004 13
2.1 Macroeconomic Stabilization Programs Adopted between 1989 and 2000....... 21
2.2 Macroeconomic Framework, 2003–07 24
2.3. International Experiences on Macroeconomic Stabilization Programs 29
3.1 Unit Cost Comparisons for Selected Countries 42
3.2 Fiscal Terms for Petroleum Exploration Contracts in Angola.................. 45
3.3 Sonangol Tax and Profit Oil Liabilities to the Government of Angola 47
3.4 A Scorecard to Assess Governance and Transparency in the Oil Sector 52
3.5 Angola's Petroleum Wealth under Different Price Scenarios................... 55

3.6	Gross Revenue Scenarios	56
3.7	Total Governance Revenues	57
3.8	Permanent Expenditure Per Capita at Different Assumptions	61
4.1	Fiscal Regimes for Diamond Mining	74
5.1	Government Policies and Behaviors and Investment Decisions	83
5.2	SADC Infrastructure Indicators	84
5.3	Trends in Merchandize Trade	88
5.4	Destination of Angolan Exports (percent)	92
5.5	Sources of Imports (percent)	92
6.1	Production of Selected Farm Products (1961–2003)	106
6.2.	Comparative Yields for Key Crops, 2003	107
7.1	Public Services According to their Perceived Importance (1 = Maximal; 7 = Minimal; %)	135
7.2	Satisfaction Rates with Public Services (1 = Maximal; 7 = Minimal; %)	136

LIST OF FIGURES

E.1	Association between Resource Concentration and Conflicts	xxxvii
E.2	Curbing Inflation (Angola: Year on Year Inflation Rate)	xxxix
E.3	Government Revenues from Oil under Different Price Scenarios and in the Absence of New Discoveries	xl
1.1	Gini Coefficients—Angola and Oil Producing Countries	9
1.2	Evolution of Angola's Real GDP Per Capita, 1960–2004	14
1.3	Composition of Domestic Expenditure, 1960–2004	15
1.4	Total Government Expenditure as a Percent of GDP for Selected Countries	16
2.1	A Snapshot of Inflation and Dollarization	22
2.2	Progress in Macroeconomic Indicators	26
2.3	Curbing Inflation	27
2.4	Foreign Exchange Interventions	28
2.5	Tradable and Nontradable Inflation Rates	31
2.6	Angola—Gross International Reserves	31
2.7	Association between Resource Concentration and Conflict	33
3.1	Oil Reserves: Selected Countries	40
3.2	Angola Oil Production	41
3.3	Total Costs for Selected Project	43
3.4	Comparison of Government Take in Selected Countries	44
3.5	Brent North Sea Oil Price Scenario	54
3.6	Gross Revenue Scenarios	56
3.7	Government Revenue Scenarios	57

3.8	Sonangol Revenues	58
3.9	Sonangol Expenditures	58
3.10	Permanent Expenditure Per Capita	60
4.1	Angola's Official Diamond Exports	68
4.2	Diamond Tax Revenues (1995–2004)	72
5.1	Importance of Informal Employment in the Urban Economy	87
5.2	Breakdown of Applied MFN Duties, 2005	90
5.3	Governance and Transparency	94
6.1	Estimated Historical Production of Major Crops, (1961–2003)	105
7.1	Annual Expenditures Per Capita in Education by Provinces	131
7.2	Annual Per Capita Expenditure in Health	132

List of Boxes

E.1	Elements of a Revenue Management Framework for Angola	xlii
3.1	Petroleum Sector Data	42
3.2	Legal and Contractual Framework	44
3.3	The Petroleum Tax Cycle in Angola	47
3.4	The Paradox and Plenty and the Case of Angola	48
3.5	The Quality of Angolan Crude Oil	54
3.6	Elements of a Revenue Management Framework for Angola	65
4.1	Certificates of Origin and the Kimberley Process	70
4.2	The International Experience on Regulating Diamond Production	75
4.3	Corporate Social Responsibility in the Diamond Sector	78
4.4	Time Frame and Actions for Diamond Sector Reform and Development	78
5.1	A Scorecard for Tackling Governance and Corruption	100
6.1	Decentralization Matters: The Mozambican "Proagri"	109
6.2	Shrinking State Intervention and Soaring Productivity—The Case of Brazil	111
7.1	Existing National Programs Addressing PRS Objectives	129

Preface

This Country Economic Memorandum (CEM) is the result of collaboration between the World Bank and the Government of Angola with the support of the British Government between 2004 and 2005. A formal discussion of the main findings and recommendations of the report was held in Luanda on May 11, 2006 with representatives of the Ministry of Planning, the Ministry of Finance, and the Central Bank.

The counterpart team in the Government of Angola was led by the Ministry of Planning under the guidance of Ms. Ana Dias Lourenço, Minister of Planning, and the leadership of Mr. Pedro Luís Fonseca, Director of Studies and Planning with the support of Messrs. Alcino Izata Conceição, Lando Teta, and Pedro Kiala. They have provided enormous support and encouragement to the CEM team and contributed useful comments and suggestions throughout the concept and implementation stages of the report. In the Ministry of Finance, especial thanks are due to Mr. José Pedro de Morais, Minister of Finance, and Messrs. Eduardo Severim de Morais and Job Graça, Vice-Ministers of Finance, for their support. Mr. Manuel Neto da Costa, Director of the Office of Studies and Foreign Relations of the Ministry of Finance (GEREI) led the discussions on behalf of the Ministry of Finance and his constant feedback was instrumental for the success of the task. The team is also grateful for the support offered by senior and technical staff in the Ministry of Agriculture, Ministry of Industry and Commerce, Ministry of Petroleum, Ministry of Geology and Mines, BNA, Sonangol, Endiama, EPAL, EDEL, and INE in the provision of the necessary information and data required to produce the CEM.

On the World Bank side, the Country Economic Memorandum was managed by Francisco Carneiro (Sr. Country Economist, AFTP1). The team included Maria Teresa Benito-Spinetto (Research Analyst, AFTP1), Fahrettin Yagci (Lead Economist, AFTP1), Nelvina Lutucuta (Economic Analyst, AFMAO), Sonia Sanchez (Consultant, AFTP1), Louise Fox (Lead Specialist, AFTPM), Charles McPherson (Sr. Advisor, COPCO), Paulo de Sá (Lead Operations Officer, LCC5C), Charles Husband (Lead Mining Specialist, COPCO), Gilberto de Barros (Sr. Private Sector Development Specialist, AFTPS), Eduardo Luis Leao de Sousa (Sr. Economist, AFTS1), Estanislao Gacituá-Marió (Sr. Social Scientist, AFTS3), Hakon Nordang (Junior Professional Associate, SDV), and Sigrun Aaslund (Operations Analyst, AFTS3). The consultants financed by the British Government through the British Embassy in Angola and DfID were Prof. Steven Kyle (Cornell University, USA), Prof. Richard Auty (University of Lancaster, UK), and Prof. Emilson Silva (Tulane University, USA).

The financial and intellectual support to the task offered by the former British Ambassador to Angola, Mr. John Thompson, renewed by his successor the current British Ambassador, Mr. Ralph Publicover, is gratefully acknowledged. Especial thanks also go to Messrs. Harry Hagan and Martin Johnston from DfID. The British Government financed background work on the impacts of phasing out fuel and price subsidies in Angola, and a Country Social Analysis (CSA). Both the subsidies study and the CSA involved a substantial effort to generate new data through a series of surveys in Luanda and other provinces in Angola and were very important inputs to the CEM and contributed to inform ongoing government policies.

Advice and comments were also received from the donor community, civil society, academia, and think tanks in Angola through a serious of consultations on specific sections of the report that were held in Luanda during preparation of the Country Economic Memorandum. Their extensive knowledge of economic and social issues in Angola was extremely valuable. Valuable comments and suggestions at different stages were also offered by Jon Shields, Alfredo Torrez, and Alexander Kyei (IMF).

The peer reviewers Jeni Klugman (Lead Economist, AFTP2), and Robert Bacon (Consultant, COPCO) provided valuable comments and suggestions. Laurence Clarke (Country Manager at the time of writing, AFMAO) and Olivier Lambert (Sr. Country Officer, AFMAO) provided invaluable support throughout the process by reading, commenting, and seeking feedback from key stakeholders in Luanda on earlier drafts of individual chapters. Peter Nicholas (Country Program Coordinator, AFCO2) contributed useful insights to several draft chapters and reviewed an earlier version of the full report. Emmanuel Akpa (Sector Manager, AFTP1) offered conceptual guidance, provided analytical advice and ensured quality control. Michael Baxter (Country Director, AFC02) read a previous draft, provided comments, and supported the whole process. Superb assistance with communication, budget management, and logistical arrangements was provided Domingas Pegado and Margarida Mendes (AFMAO). Ligia Murphy (AFTP1) showed outstanding commitment and provided impeccable assistance in formatting and editing the report.

Vice President	:	Obiageli Katryn Ezekwesili
Country Director	:	Michael Baxter
Country Program Coordinator	:	Peter Nicholas
Country Manager	:	Alberto Chueca-Mora
Sector Manager	:	Emmanuel Akpa
Task Team Leader	:	Francisco Galrão Carneiro

Acronyms and Abbreviations

AfDB	African Development Bank
AGOA	Africa Growth and Opportunity Act
ANIP	Agência Nacional de Investimentos Privados
ASCORP	Angola Selling Corporation
AUPEC	Aberdeen University Petroleum Economics Consultancy
BNA	Banco Nacional de Angola
CABOG	Cabinda Gulf Oil Company
CPI	Consumer price index
DFID	UK Department for International Development
DNI	Directoria Nacional de Impostos
EBA	Everything But Arms
ECP	Estratégia de Combate à Pobreza
EDEL	Empresa de Distribuição de Electricidade de Luanda
EITI	Extractive Industries Transparency Initiative
Endiama	Empresa de Diamantes de Angola
ENE	Empresa Nacional de Electricidade
EPA	Economic Partnership Agreement
EPAL	Empresa Pública de Águas de Luanda
FAO	Food and Agriculture Organization
FAS	Fundo de Acção Social
FDES	Fundo de Desenvolvimento Económico e Social
FDI	Foreign direct investment
FLEC	Frente para a Libertação do Enclave de Cabinda
FNLA	Frente Nacional de Libertação de Angola
GARE	Gabinete de Redimensionamento Empresarial
GDP	Gross domestic product
GoA	Government of Angola
HDI	Human Development Indicators
IDR	Inquérito de Despesas e Receitas
IFI	International Financial Institutions
IMF	International Monetary Fund
INE	Instituto Nacional de Estatísticas
Kz	Angolan Kwanza (currency unit)
LPG	Liquified petroleum gas
LUPP	Luanda Urban Poverty Program
MBD	Million barrels per day
MFN	Most favored nation
MICS	Multiple Indicator Cluster Survey
MINADER	Ministry of Agriculture and Rural Development
MOF	Ministry of Finance
MOP	Ministry of Petroleum
MPLA	Movimento Popular para a Libertação de Angola

MTFF	Medium-Term Fiscal Framework
NEPAD	The New Partnership for Africa's Development
OECD	Organization for Economic Development and Cooperation
OPEC	Organization of Petroleum Exporting Countries
PDV	Present Discounted Value
PIP	Public investment program
PIT	Petroleum income tax
PSA	Production Sharing Agreement
PTT	Petroleum transaction tax
SADC	Southern African Development Community
SIGFE	Sistema Integrado para a Gestão das Finanças do Estado
SODIAM	Sociedade de Comercialização de Diamantes
SSA	Sub-saharan Africa
UNDP	United Nations Development Program
UNITA	União Nacional para a Independência Total de Angola
UXO	Unexploded ordinance
WHO	World Health Organization

Currency Equivalents
Currency Unit: Angolan Kwanza (Kz)
US$ = Kz80.16892 (Exchange Rate Effective as of October 2, 2006)
Weights and Measures
Metric System

Fiscal Year of Budget: January 1–December 31

Prefácio

Esta versão do Memorando Económico para o País sobre Angola incorpora comentários e sugestões recebidas das autoridades angolanas durante a reunião para apresentação e discussão do relatório, que aconteceu no dia 11 de Maio de 2006, no Ministério do Planeamento, em Luanda, com a presença de representantes do Ministério do Planeamento, Ministério das Finanças, e do Banco Nacional de Angola. O Banco Mundial recebeu ainda comentários por escrito enviados pelo Gabinete de Estudos e Relações Internacionais do Ministério das Finanças.

Um breve resumo dos principais tópicos discutidos durante a reunião do dia 11 de maio bem como dos assuntos ventilados nos comentários do Ministério das Finanças é apresentado a seguir.

Constrangimentos Externos: A existência de constrangimentos externos, alheios ao controlo do governo, que podem afectar o desempenho da economia e das políticas económicas, sendo que o principal destes constrangimentos é o próprio preço do petróleo.

Papel do Sector Agrícola: Enquanto reconhece-se que o sector agrícola tem papel fundamental na economia, há uma limitação para avaliar-se qual é exatamente o grau de competitividade do sector devido à falta de uma noção clara sobre a estrutura de custos de produção. Para além disso, o sector enfrenta outros constrangimentos tais como a qualidade das terras, que impede a reprodução das culturas produzidas na era colonial.

Apoio à Agricultura: Quanto ao apoio a prestar à agricultura, as autoridades comentaram que o MINADER e os Governos provinciais deveriam prestar o apoio institucional e os serviços necessários. Mas adicionaram que será necessário incitar as empresas agro-industriais e de exportação a instalarem-se em determinadas áreas. Os financiamentos aos pequenos produtores deveriam ser fornecidos através de bancos/instituições locais de micro-crédito com o objectivo de introduzir uma certa competitividade nesse sector.

Indústria Transformadora: A actual configuração da balança de pagamentos de Angola implica por si só na valorização da moeda. A moeda valorizada funciona como um forte constrangimento para a competitividade da indústria local. Esta situação gera a necessidade de se procurar reduzir os custos de produção em relação aos países vizinhos.

Apreciação da Taxa de Câmbio: Foi ressaltada a aparente contradição entre a necessidade de se reduzir a inflação, o que demandaria evitar tentar combater a tendência à apreciação da taxa de câmbio em face do volume de divisas que será acumulado em anos futuros, e a necessidade de se diversificar a economia e se promover um aumento da competitividade dos sectores não-minerais. Diante disso, ressaltou-se que há a necessidade de se procurar reduzir os custos domésticos através da reparação das infraestruturas e da identificação de outros factores constrangedores que possam ser eliminados, assumindo o constrangimento de uma taxa de câmbio apreciada como inevitável em anos futuros.

Investimento na Capacitação: As autoridades enfatizaram que o governo tem estado a investir significativamente em capacitação nas mais diversas areas, uma vez que reconhece-se que o sistema de ensino actual não contribui de maneira efectiva para gerar mão de obra qualificada com a urgência que se precisa.

Transição para a Democracia: Na apresentação da principal mensagem do memorando, o documento justifica a necessidade de completar a transição para a democracia multipartidária com a necessidade de acomodar as tensões étnicas. As autoridades sugeriram que seria necessário mencionar que a democracia ajudaria ainda a amenizar tensões sociais.

Composição das Despesas Públicas: O governo enfatizou que a adopção de preços conservadores para o barril do petróleo na elaboração do orçamento representa uma boa opção. Com base nesta opção, portanto, dever-se-á ter um limite de despesa pública baseada nas perspectivas de médio prazo das receitas avaliadas com base nos preços projectados de longo prazo. No entanto, as autoridades sugerem que os excedentes constituiriam reservas que deveriam ser aplicadas em mercados/investimentos internacionais que serviriam de colateral para eventuais empréstimos destinados a financiar os investimentos internos. Desse modo, os juros sobre os empréstimos seriam compensados pelos juros e dividendos das aplicações internacionais.

Reforço do Sector Privado: O fortalecimento do sector privado requer que o Estado, ao mesmo tempo que deva se preocupar pelas infraestruturas, governança e ambiente de negócios, deva também contemplar, nas suas políticas de compensação dos custos de ajustamento, a requalificação da mão-de-obra, pois o investimento privado tem tendência a exigir cada vez maiores qualificações da mão-de-obra. Fazer passar as pessoas de não qualificadas para semi-qualificadas, a qualificadas e a altamente qualificadas exige um empenho do governo, particularmente nos países em desenvolvimento como Angola. A decisão de investimento privado dependerá muito disso também.

Outros assuntos que foram objecto de discussão envolveram (i) a questão da legislação do petróleo (que não mudou e não alterou, portanto, contractos existentes); e (ii) a necessidade de se gerar informação confiável sobre custos unitários nas áreas da educação e da saúde para aumentar o grau de eficiência dos investimentos neste sectores.

Em suma, o Memorando foi considerado como um conjunto de considerações e recomendações gerais que procuram ajudar o Governo na definição e implementação de políticas de modo a que as receitas petrolíferas sejam bem aproveitadas.

Sumário Executivo

As Principais Mensagens do Relatório

Este *Memorando Económico do País* identifica seis áreas centrais para as quais é necessário um plano de abordagem para o desenvolvimento de uma estratégia de crescimento de base ampla. Para a concepção de uma estratégia de desenvolvimento sustentável para o país, as autoridades vão precisar de contemplar um certo número de questões em seis áreas principais que foram consideradas como constrangimentos ao crescimento e ao desenvolvimento equitativo em Angola: (i) transição incompleta para uma economia de mercado; (ii) gestão macroeconómica; (iii) governação e transparência na gestão da riqueza mineral; (iv) clima de negócios; (v) agricultura; e (vi) prestação de serviços públicos aos pobres. Resume-se, abaixo, a abordagem estratégica recomendada para cada uma destas áreas. As secções subsequentes deste susmário executivo dão relevo às conclusões de carácter mais amplo e às recomendações principais do *Memorando Económico do País*.

- Em primeiro lugar, Angola precisa de concluir a transição para uma economia de mercado. Com o advento da Independência em 1975, o governo angolano optou por um sistema económico centralizado que, hoje, se julga ter contribuído, a par da crescente dependência do petróleo, para constranger o desenvolvimento de instituições sólidas que promovam o aparecimento de um sector privado dinâmico. A transição para um completo sistema de economia de mercado irá exigir comprometimento político aos níveis mais altos visto que terão que ser desmantelados os arreigados direitos adquiridos. Adicionalmente, o país também vai ter que completar a transição para uma democracia multipartidária, o que irá ajudar a acomodar as tensões étnicas que, no passado, permearam e atearam o conflito civil em Angola e que continuam latentes no presente. Estes pontos são tratados nos Capítulos 1 e 2.
- Em segundo lugar, na frente macroeconómica, ainda precisam de ser resolvidas as contínuas deficiências na concepção e execução de políticas, especialmente ao nível agregado. O progresso recente na gestão da economia é encorajador, sobretudo o sucesso conseguido na redução da inflação e reforço da posição fiscal do governo. No entanto, estes factos têm sido amplamente influenciados por desenvolvimentos bastante favoráveis no sector petrolífero. Para garantir a sustentabilidade da recém conquistada estabilização macroeconómica, Angola precisa de adoptar uma combinação sólida de políticas económicas, centradas na melhor gestão das despesas públicas e administrar algumas tensões que são comuns a economias em fase de transição da guerra para a paz e de uma economia dirigida para uma economia de mercado. Os Capítulos 2 e 3 apresentam algumas opções de políticas macroeconómicas.
- Em terceiro lugar, há que definir uma estratégia clara para se gerir a crescente riqueza do país em petróleo e diamantes com base numa governação salutar e em princípios de transparência. Nas próximas duas décadas, a economia angolana vai registar um benefício excepcional de volumosas receitas petrolíferas com uma concomitante melhoria fiscal. Prevê-se que a produção de diamantes também tenha

um crescimento robusto. Os lucros decorrentes da maior produção representam, contudo, a redução das reservas de petróleo e diamantes do país. Isto levanta a questão de como utilizar as receitas dos recursos não renováveis de modo a criar fontes de rendimento para o futuro, incluindo para as gerações vindouras. Uma governação sólida e princípios de transparência deverão guiar o estabelecimento de qualquer estratégia de gestão de recursos naturais em Angola. Nos Capítulos 3 e 4 discutem-se em pormenor as questões e opções nesta área.

- Em quarto lugar, e relacionado com o ponto acima, há uma necessidade de se melhorar o ambiente de negócios e o clima de investimento em Angola. O ambiente de negócios no país é um dos menos favoráveis do mundo inteiro. Se as autoridades quiserem promover uma recuperação económica de base ampla do país, com mais empregos e rendimentos mais elevados para o angolano médio, então há que implementar de imediato medidas em prol da actividade económica que permitam às empresas competir mais eficazmente numa economia aberta. De outra forma, a economia angolana continuar a estar altamente dependente da exploração dos recursos minerais e a maioria da população continuará a não ser beneficiada pela riqueza mineral. O Capítulo 5 centra-se nos desafios e oportunidades associados com o desenvolvimento do sector privado não mineral em Angola.

- Em quinto lugar, considerando as áreas com potencialidade fora dos sectores minerais, a importância da agricultura como uma fonte de emprego e de rendimentos não deverá ser negligenciada. Angola é um produtor natural de cultivos alimentares e comerciais e a agricultura tem potencial para contribuir para o aumento do crescimento económico nos anos que se avizinham. As políticas governamentais deveriam apoiar a proliferação de pequenos agricultores mas, ao mesmo tempo, promover um ambiente propício que estimule os investimentos no sector comercial privado. No Capítulo 6 esboça-se uma estratégia para aumentar a produção agrícola e melhorar os incentivos que possam contribuir para uma maior competitividade no sector.

- Por último, como parte do dividendo da paz a ser repartido com a população angolana, tem que se melhorar a qualidade e a prestação de serviços públicos, especialmente os que se destinam aos pobres. Sendo um país em fase de pós conflito, a melhoria do bem-estar social da população, incluindo a mais pobre, representa um enorme desafio para Angola. O país tem uma boa probabilidade de responder a este desafio graças aos crescentes fundos disponíveis com a exploração dos seus recursos naturais. No Capítulo 7 discutem-se as opções viáveis para se melhorar o bem-estar social da maioria da população de Angola.

O Caminho a Percorrer Envolve Riscos

O caminho a seguir envolve riscos políticos de resolução difícil. Provas empíricas amplas têm revelado que a dependência de recursos naturais é particularmente problemática, uma vez que pode ser facilmente capturado pelas elites, eliminando o incentivo de o Governo envolver activamente todos os cidadãos. Isto destrói tanto a capacidade como a legitimidade do Estado, exacerbando divisões sociais e levando até ao conflito directo quanto aos recursos propriamente ditos. A pesquisa do Banco Mundial e de outras fontes refere que a dependência de recursos é uma das principais causas das guerras civis (ver Figura E.1).[1] No caso de Angola, avançar com a agenda de reformas nas áreas de governação, transparência, gestão das finanças públicas, ambiente de negócios e prestação de serviços públicos vai ser um desafio político mas os riscos de não se adoptarem nenhumas reformas ou de se esperar demasiado para se actuar pode ter um efeito devastador face às enormes lacunas que o país precisa de preencher. Não obstante, se bem que um aumento das receitas de recursos naturais tenha potencial para acentuar uma atitude de procura de lucros ilegítimos e dar aos governos uma desculpa para se adiar a reforma, pode também fazer com que os governos interessados em promover as reformas se empenhem em executá-las. Angola tem agora uma verdadeira oportunidade para utilizar a sua riqueza em recursos naturais para mitigar as tensões e adoptar as medidas difíceis que podem por a economia no caminho do desenvolvimento sustentável. O Banco Mundial está pronto a apoiar o Governo nesse esforço.

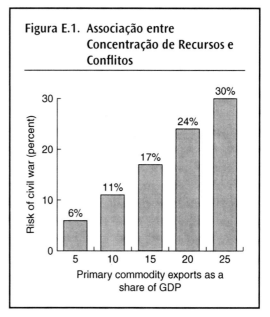

Figura E.1. Associação entre Concentração de Recursos e Conflitos

Fonte: Bannon and Collier (2003).

Realidades Socioeconómicas

Um Produtor de Petróleo e de Diamantes de Primeiro Nível

Tendo sido em tempos um dos maiores produtores de cultivos alimentares, Angola é hoje em dia conhecida como um dos grandes produtores de petróleo. Antes da independência, Angola era mais conhecida como um exportador de café do que como um país exportador de petróleo. Angola era não apenas o quarto maior exportador de café do mundo, como também exportava mais de 400 000 toneladas métricas de milho por ano, sendo um dos maiores exportadores de artigos alimentares da África Subsaariana. A estrutura da economia sofreu então uma grande mudança, depois de 1973, com o aumento substancial da quota dos sectores de mineração e de serviços na composição do PIB. O país é hoje o segundo maior produtor de petróleo de África.

Angola também é o quarto produtor mundial de diamantes brutos em termos de valor, com potencial para se tornar um dos maiores produtores universais de diamantes. O país detém uma quota de cerca de 12% do mercado mundial e, uma grande parte da sua produção,

---

1. Movimentos separatistas violentos podem frequentemente estar associados ao petróleo. Entre os exemplos contam-se Aceh na Indonésia, Biafra na Nigéria, conflitos no Sudão, Chade, Congo Brazzaville e mesmo em Angola (ver Busby et. al., 2002).

tem qualidade de pedra preciosa. Em 2000, as reservas de diamantes foram calculadas em 40 milhões de quilates em depósitos aluviais e em 50 milhões de quilates em jazidas quimberlíticas que só agora começaram a ser explorados. Conhecem-se cerca de 700 quimberlíticas de dimensão variável (10-190 hectares) em Angola, alinhadas com uma tendência no sentido SW/NE ao longo do país e que se prolongam pela República Democrática do Congo.

Para além do petróleo e dos diamantes, o país está generosamente dotado de recursos agrícolas que, na sua maioria, continuam intactos. Os cultivos alimentares vão da *mandioca* na zona húmida do norte e do nordeste ao *milho* nas terras altas do centro e ao sorgo/milho-miúdo nas províncias mais secas do sul. As *batatas* são um cultivo importante no planalto central e o *arroz* também é cultivado em grande dimensão no norte. Faz-se a criação de *gado* em grandes áreas do planalto central mas é especialmente importante nas províncias do sul do Cunene, Huila e Namibe onde se calcula que existam 3 milhões de cabeças de gado. O *café*, o cultivo comercial por excelência durante os tempos coloniais, dá-se bem nas terras altas do Uíge e Malange, em todo o Kuanza Norte e estende-se até ao Huambo e Bié. Também é importante notar que cerca de dois terços da população de Angola vive em zonas rurais e fazem da agricultura o seu modo de vida, o que actualmente representa menos de 10% do PIB e recebe menos de 1% do total das despesas orçamentais.

Indicadores de Pobreza e de Bem-estar Assustadores

Apesar da significativa riqueza natural do país, os indicadores sociais existentes ainda reflectem os baixos padrões de vida. De acordo quer com o IDR 2001 (Inquérito das Receitas e Despesas) quer com o MICS 2002 (Inquérido de Grupos de Indicadores Múltiplos), aproximadamente 70% da população vive com menos de 2 dólares por dia e a maioria dos angolanos ainda não tem acesso aos cuidados de saúde básicos. Cerca de uma em cada quatro crianças angolanas morre antes dos cinco anos, 90% das quais sucumbem à malária, diarréia ou infecções do aparelho respiratório; a taxa de mortalidade maternal é uma das mais altas da SSA (1 800 por 100 000 nascimentos); e três em cada cinco pessoas não têm acesso a água potável ou saneamento. De acordo com as estatísticas oficiais, a taxa de prevalência de VIH/SIDA é relativamente baixa, afectando aproximadamente 3,9% de adultos.[2] Relativamente ao ensino, a taxa de matrículas no ensino primário é muito baixa, situando-se em 56%, e é afectada por entradas tardias no ensino, taxas de repetição e de abandono escolar elevadas. Aproximadamente 33% da população ainda é analfabeta, embora nas zonas rurais chegue a atingir 50%.

Um Ambiente de Negócios Pouco Convidativo

O ambiente de negócios em Angola é problemático. Segundo o levantamento do Banco Mundial, intitulado 2006 *Doing Business*, o estabelecimento de uma empresa em Angola leva em média cerca de 146 dias, mais do que o dobro da média na região. A obtenção de alvarás é um procedimento moroso e caro. Calcula-se em 326 dias o tempo necessário para cumprir todas as formalidades de licenças e alvarás. O registo de propriedade também demora cerca de 11 meses e custa mais do que 11% do valor da propriedade. Quando se

---

2. Estimativas baixas apontam para 1,9% e as estimativas elevadas indicam 9,4% (UNAIDS, 2004).

compara com outros países, não existe uma grande protecção do investidor e, a obtenção de crédito é igualmente difícil. Para uma importação média são precisos 10 documentos, 28 assinaturas e 64 dias. A execução de um contrato também é complicada—leva cerca de 1011 dias civis a resolver uma disputa, a partir do momento que um queixoso apresenta queixa em tribunal até à sua resolução ou pagamento. A resolução de disputas nos países vizinhos de Angola, seus potenciais concorrentes, demora muito menos tempo: 274 dias na Zâmbia; 270 na Namíbia e 154 no Botswana.

Para além de um ambiente de negócios pouco favorável, Angola está entre os países com os índices mais elevados de percepção de corrupção. Nos rankings estabelecidos pela Transparency International quanto ao Índice de Percepções de Corrupção (IPC), os países em desenvolvimento exportadores de petróleo encontram-se no terço final dos países referidos. Na publicação mais recente do IPC, Angola ocupava o 151° lugar, o nono a contar do último. Isto é particularmente grave porque se reconhece a corrupção como um dos factores individuais que mais inibe o investimento do sector privado e o crescimento.

Agenda para Reforma Macroeconómica

Progresso Tangível e Oportunidades Promissoras

Mais recentemente, os esforços do governo para reduzir a inflação têm sido bem sucedidos. Entre 1999 e o acordo de paz de 2002, a inflação anual dos preços ao consumidor baixou de cerca de 300% para aproximadamente 100% (ver Figura E.2). Após a adopção de um programa de estabilização, em Setembro de 2003, a inflação baixou consideravelmente de novo e, em Dezembro de 2004, a taxa de inflação relativa ao período de 12 meses tinha caído para 31%. A melhoria foi sobretudo devida ao facto de o Governo ter evitado

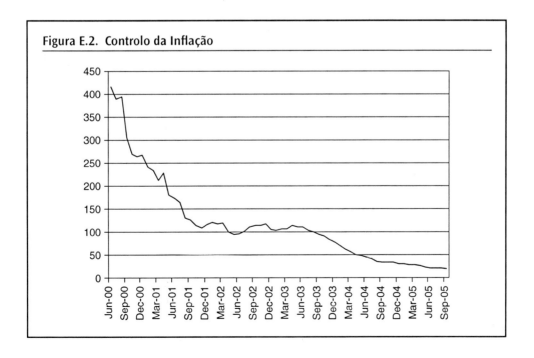

Figura E.2. Controlo da Inflação

recorrer à emissão de moeda para financiamento do défice a par de défices fiscais menos elevados. Em 2005, a taxa de inflação cumulativa baixou para 18,5% e a projecção para 2006 é de uma taxa anual de 10%.

Não obstante o progresso recente, deveria prestar-se mais atenção às contínuas deficiências na concepção e execução de políticas. É preciso reforçar a estabilização até agora conseguida com uma melhor coordenação entre a política fiscal e as políticas monetárias e cambiais. Estas políticas têm que definir uma estratégia consistente para se absorver as próximas receitas extraordinárias do petróleo sem se inibir o crescimento fora dos sectores minerais. Um primeiro passo na direcção certa seria a redução das despesas públicas com o consumo e o aumento das despesas nos investimentos produtivos (e.g., infra-estruturas) e outros do género que tenham por efeito o abaixamento dos custos domésticos de toda a economia. Em certa medida, esta estratégia já está a ser seguida pelas autoridades, mas existem certas áreas em que são necessárias medidas complementares, as quais são discutidas abaixo.

Alinhar a Política Económica com as Realidades Macroeconómicas

Deveriam adoptar-se instrumentos para suavizar o impacto dos ciclos petrolíferos. A combinação das políticas fiscal, financeira, comercial e monetária tem de atender à necessidade de se estimularem os investimentos privados nos sectores do comércio e de se aumentar a dimensão e a sofisticação dos sectores não extractivos. Tal exigirá (i) o controlo da procura agregada (i.e. a rápida expansão dos gastos ou crédito públicos como um resultado indirecto do *boom* do petróleo); e (ii) o uso de um enquadramento económico a médio prazo para guiar as decisões da política económica. Nos instrumentos destinados a suavizar o impacto dos ciclos petrolíferos na economia e a criar um ritmo desejável da industrialização não petrolífera contam-se:

- *Um envelope fiscal prudente para os anos seguintes.* É importante ter presente o perfil em forma de bossa das receitas fiscais relativas ao petróleo e o facto de que baixarão rapidamente durante a década seguinte, caso não haja novas descobertas (ver Figura E.3). Serão precisos excedentes substanciais quando se realizar a rápida extracção de petróleo durante o resto desta década e grande parte da que se segue. Mas isto pode não ser politicamente possível. Adicionalmente, podem não se concretizar as expectativas quanto ao crescimento futuro do sector não petrolífero, enquanto é incerto o valor actual descontado (VAD) da riqueza petrolífera futura e sujeito à volatilidade do preço do petróleo. Assim, embora haja uma ampla margem para aumentos das despesas públicas, as preocupações referidas acima requerem um envelope fiscal conservador.
- *Um enquadramento económico de médio prazo.* Actualmente, as decisões que afectam a política macroeconómica são tomadas num horizonte máximo de dois anos, visto que os programas de despesas dos orçamentos anuais são determinados apenas pelas projecções de receitas para o ano seguinte. Apesar do facto de o governo ter estado a adoptar uma abordagem conservadora com vista à indicação do preço do petróleo, com base no qual se fundamentam as projecções das receitas, esta abordagem pode dar origem a ciclos abruptos no domínio das despesas, situação que se pode tornar insustentável a longo prazo. Seria preferível uma abordagem que assentasse

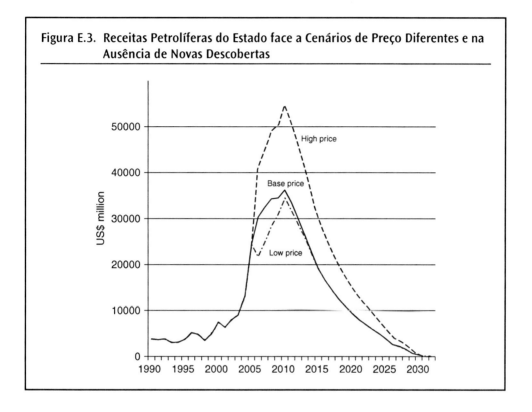

Figura E.3. Receitas Petrolíferas do Estado face a Cenários de Preço Diferentes e na Ausência de Novas Descobertas

em tectos anuais para as despesas públicas sobre as perspectivas de receitas a médio prazo, avaliadas num preço do petróleo a longo prazo. No âmbito desta abordagem, a recomendação feita no parágrafo anterior de se gerarem excedentes fiscais consistentes poderia, com tempo, ser alcançada de forma a que se pudessem acumular reservas financeiras durante os anos em que a produção de petróleo atingisse o seu cume com vista a sustentar as despesas quando as reservas de petróleo vierem, eventualmente, a cair.

- *Políticas fiscais de combate ao carácter cíclico dos preços do petróleo.* O carácter pró cíclico das despesas fiscais e dos preços do petróleo é perigoso e pode transmitir volatilidade ao resto da economia. Um objectivo chave das políticas para Angola deveria ser a observância de estratégias fiscais destinadas a por termo à resposta pró cíclica das despesas à volatilidade dos preços do petróleo. Neste sentido, é louvável a abordagem presentemente utilizada pelas autoridades de adoptarem um preço conservador para o petróleo na preparação do orçamento, mas com a anotação de que deveria ser complementada com as recomendações dos parágrafos anteriores.
- *Melhorias rápidas e audaciosas nas práticas de aquisições.* Em termos simples, o nível de despesas deveria ser determinado levando em conta a sua provável qualidade e a capacidade de a administração as executarem eficientemente. Neste domínio, um alargamento brusco dos programas de despesas associado com os benefícios extraordinários do petróleo implica riscos graves. Um programa de

despesas públicas precipitado pode exceder a capacidade de planeamento, implementação e de gestão do Governo, podendo daí advir a dificuldade de se impedir despesas esbanjadoras.

Opções Institucionais para se Gerirem as Receitas Extraordinárias

Será necessário um enquadramento institucional para se gerir eficazmente a riqueza petrolífera de Angola. Independentemente de Angola adoptar ou não as recomendações indicadas acima, o Governo vai ter que criar a capacidade institucional exigida para se gerirem receitas futuras. O anúncio recente da abertura de uma conta de reserva do petróleo no Banco Central onde ficariam acumuladas parte das receitas extraordinárias do petróleo e a utilização pelo Ministério das Finanças de um modelo de projecção das receitas petrolíferas são acontecimentos dignos de registo, mas que precisam de ser complementados por medidas adicionais.

A definição de regras para a conta de reserva do petróleo deverá apoiar-se nos sucessos e nos fracassos de experiências anteriores noutros países produtores de petróleo. Como o Governo de Angola está a considerar a criação de uma conta de reserva do petróleo, seria talvez útil considerar, para o futuro, os aspectos seguintes que se baseiam noutros modelos que estão actualmente a ser desenvolvidos e implementados noutros países (na caixa E.1 abaixo apresenta-se um sumário destes aspectos):

- A conta de reserva do petróleo deveria ser o receptáculo escolhido para depósito das receitas extraordinárias do petróleo. A captação parcial das receitas do petróleo por parte dos vários ministérios ou empresas públicas seria um grande obstáculo à gestão eficaz e sustentável das receitas.
- Deveria existir um tecto, previamente determinado, que limitasse o montante de recursos acumulados na conta de reserva do petróleo para se evitarem problemas de economia política. É sensato reconhecer antecipadamente que vão exercer-se pressões crescentes para se dobrarem as regras e levantar mais recursos se estes se tornarem demasiado volumosos. Um limite definido antecipadamente e acompanhado de regras precisas quanto ao modo como serão gastos os fundos não destinados a poupança evitaria problemas no domínio da economia política.
- As transferências para o orçamento deveriam basear-se na regra acordada sobre poupança/despesas. Sem regras claras, existe o risco de as transferências se tornarem cada vez mais dependentes de ciclos económicos e políticos de curto prazo, o que poderia minar perigosamente a função da conta.
- O BNA deveria ser designado como o gestor das operações (do dia-a-dia) da conta. Para uma gestão eficiente, transparente e cumpridora das regras estabelecidas é necessário dispor de funcionários imparciais no Banco Central.
- As regras para o investimento dos activos da conta (por exemplo instrumentos de baixo risco do tesouro estrangeiro) têm de ser claras, acordadas e publicadas.
- Terá de ser criado um comité de supervisão de alto nível (ministérios chave mais conselheiros externos qualificados).
- Terão de ser feitas auditorias profissionais anuais da conta de reserva do petróleo e de ser publicado o relatório do auditor.

> **Caixa E.1: Elementos de um Quadro de Gestão das Receitas para Angola**
>
> **Consolidação e cobrança das receitas**
> – Todas as receitas cobradas referentes ao petróleo são consolidadas através da conta de reserva do petróleo
> – As receitas são contabilizadas de acordo com normas de contabilidade acordadas e transparentes
> – As receitas são publicadas de uma forma acessível e pontual
>
> **Definição de poupança e consumo**
> – As transferências da conta de reserva do petróleo para o orçamento baseiam-se nas regras acordadas sobre poupança/despesas.
> – As regras são claras, previsíveis e públicas e a sua aplicação não depende de discrição administrativa ou política
> – As preocupações macroeconómicas e de sustentabilidade são da maior importância na elaboração das normas sobre poupança e consumo
>
> **Institucionalização do mecanismo de transferências**
> – As transferências do organismo que efectua a cobrança das receitas para o orçamento e para a conta de reserva do petróleo obedecem a regras previamente estabelecidas e processam-se automaticamente, independentemente da vontade administrativa ou política
>
> **Gestão da Conta**
> – A gestão dos fundos baseia-se em regras claras, transparentes, acordadas e previsíveis que atribuem responsabilidades e necessidades de reporte claras.
> – O BNA é designado o gestor das operações (do dia-a-dia) da conta.
> – As regras para o investimento dos activos da conta têm de ser bem claras, acordadas e publicadas.
> – Os activos do fundo deveriam, em grande medida, ser investidos no estrangeiro e em instrumentos seguros
> – O BNA informa sobre o desempenho da conta de reserva do petróleo e a afectação dos activos segundo um calendário previamente definido e os relatórios são tornados públicos de uma forma acessível
> – Tem que ser formado um comité de supervisão de alto nível (ministérios chave mais conselheiros externos qualificados).
> – A conta de reserva do petróleo, sua gestão e normas têm de obedecer a requisitos de transparência rigorosos.
> – Deveria existir um tecto previamente definido quanto ao limite dos recursos acumulados para se evitar problemas de economia política.

▪ A conta de reserva do petróleo, sua gestão e normas têm de obedecer a requisitos rigorosos de transparência.

Reforçar a Governação e a Transparência

A gestão apropriada da riqueza petrolífera e de diamantes em Angola irá exigir melhorias nas áreas de governação, transparência e capacidade institucional. Numa economia fortemente dependente dos recursos minerais, a presença de instituições débeis pode convidar à procura de recompensas monetárias ilícitas e ao desperdício. No caso de Angola, onde o nível da qualidade institucional é baixo, em parte por causa dos efeitos da guerra e em parte por problemas associados com o chamado Paradoxo da Abundância, é absolutamente necessário melhorar os factores que podem afectar a qualidade da governação. Para tal, serão precisos esforços adicionais nas 4 áreas de vasto alcance que se enumeram a seguir.

Lidar com os Conflitos de Interesse

A relação entre a Sonangol, a empresa petrolífera nacional, e o Governo gera conflitos de interesse e opõe-se à boa prática na gestão das finanças públicas. A Sonangol desempenha papéis múltiplos vis à vis o Governo, incluindo actividades que normalmente pertenceriam ao Tesouro e ao Banco Central. É um contribuinte, efectua actividades parafiscais, faz o investimento de fundos públicos e, como concessionária, é um regulador do sector. Este programa de trabalho multifacetado cria conflitos de interesse e caracteriza uma relação complexa entre a Sonangol e o Governo que provoca o enfraquecimento do processo orçamental formal e cria incerteza no que toca à posição fiscal do Estado.

Existe uma preocupação semelhante quanto às operações da Endiama, a companhia nacional de diamantes. O Ministério da Geologia e Minas é responsável pela implementação do enquadramento legal e regulador do sector, pela emissão de direitos minerais e pelo levantamento geológico, enquanto o mandato para aprovar as concessões quimberlíticas pertence ao Conselho de Ministros. A Endiama é, por lei, o maior accionista em todos os novos empreendimentos de diamantes e possui funções reguladoras sobre a selecção de companhias a quem vão ser atribuídos os novos direitos minerais sobre diamantes, a negociação de contratos de mineração e a monitorização e controlo de actividades das empresas de diamantes. A Endiama é, portanto, ao mesmo tempo, um operador e uma entidade reguladora da indústria de diamantes.

Logo, são necessárias reformas institucionais nos sectores do petróleo e dos diamantes. No sector do petróleo, o papel da Sonangol deveria ser reavaliado com vista a eliminar o conflito de interesses e a melhorar a qualidade e a eficácia da gestão das finanças públicas em Angola. No sector dos diamantes, será preciso introduzir uma estrutura transparente de concessão de licenças, da entidade reguladora e tributária que permita evitar os conflitos de interesse e o acesso privilegiado aos direitos de desenvolvimento. Estas questões são discutidas individualmente abaixo.

Reavaliar o Papel da Sonangol

Mais tem que ser feito para se acabar com o papel duplo da Sonangol. Embora tenha havido iniciativas para aumentar a transparência nas relações entre a Sonangol e o Governo, os procedimentos actuais apresentam várias desvantagens, sem contar com o registo de actividades fiscais fora do orçamento, onde estas pertencem: a) como se disse acima, não é transparente; b) conduz a disputas frequentes; e c) falseia a posição da Sonangol em termos de impostos. O Governo de Angola está agora a tomar medidas com vista a fazer reflectir no orçamento as actividades parafiscais e a Sonangol está a actuar para que a sua auditoria externa identifique e audite todas essas actividades realizadas em substituição do Governo.

Registam-se, com agrado, estas duas medidas. No entanto, a prática corrente ainda funciona bastante como um mecanismo de resistência e estão por concretizar os esforços para a sua eliminação.

A Sonangol pode contribuir mais eficazmente para o desenvolvimento do país se se concentrar na sua actividade central. É um facto reconhecido que as actividades desempenhadas pela Sonangol em representação do Governo criam um ónus administrativo e operacional na companhia. Se estas forem abandonadas e a Sonangol se centrar exclusivamente no seu negócio principal, a companhia poderia contribuir ainda mais para o desenvolvimento do país. Referem-se alguns exemplos de áreas onde uma maior intervenção da Sonangol poderia fazer diferença: (a) investimentos para aumentar as capacidades de armazenamento; (b) investimentos para aumentar a distribuição de combustível ao interior do país; (c) formação de mão-de-obra angolana para trabalhar nas IOC; e (d) investimentos em projectos sociais.

Resolver as Questões de Governação e de Transparência

Houve algum progresso na melhoria da governação e transparência no sector petrolífero desde 2002. Foram adoptadas várias recomendações feitas anteriormente pelo Banco, quer no *Estudo de Diagnóstico do Petróleo* quer no PEMFAR e que se resumem a seguir:

- O governo tem vindo a publicar pormenores dos pagamentos de petróleo recebidos (por bloco, por tipo de pagamento, com resumos anuais por companhia) na página da Internet do Ministério das Finanças, se bem que com um atraso significativo de 6 meses em média.
- Estão a ser feitas auditorias do sector petrolífero. As posições financeiras das empresas do grupo Sonangol foram sujeitas, pela primeira vez, a uma auditoria extensiva realizada por Ernst & Young em 2003. A auditoria relativa a 2004 ficou recentemente concluída. Contudo, os relatórios dos auditores ainda não foram publicados.
- As auditorias recentes observaram regras de contabilidade aceitáveis. Foi aplicada, sempre que pertinente, a legislação sobre contabilidade recentemente emitida em Angola às auditorias referidas acima e, quando esta nova legislação não fosse aplicável, utilizaram-se as regras IAS. Espera-se que, antes do fim de 2006, a Sonangol tenha adoptado integralmente os padrões IAS.
- A padronização do SIGFE está a progredir constantemente. Tem havido um progresso constante na implementação e na padronização integral do SIGFE angolano e, actualmente, a maior parte das despesas do governo está registada no sistema. De referir que o progresso no domínio da inclusão no sistema de dados relativos a receitas permanece muito ténue.
- O governo está a utilizar um modelo de projecção das receitas petrolíferas. O Ministério das Finanças assinou um contrato com a AUPEC, Aberdeen University Petroleum Economic Consultancy, com vista a implementar um modelo de previsão das receitas petrolíferas e dar assessoria à DNI quanto ao modo de reforçar a sua capacidade no domínio da tributação do petróleo.

Mas o progresso tem sido lento na resolução de outras questões estruturais de gestão financeira igualmente importantes. O Governo começou a reforçar a capacidade do

Ministério das Finanças para o controlo das despesas com a padronização do SIGFE e a criação de uma unidade de programação fiscal. Está também em vigor uma iniciativa com o intuito de se conseguir alguma supervisão sobre as operações efectuadas pela Sonangol em representação do Tesouro, que se traduz no registo à posteriori dessas operações no SIGFE. Mas mais tem que ser feito para se chegar ao ponto desejado da sua "normalização". Entre outras melhorias que se sugerem para a gestão das finanças públicas incluem-se:

- *Operações parafiscais da Sonangol:* actualmente, as operações parafiscais da Sonangol estão reflectidas no orçamento (com um atraso) mas não há nenhuma indicação clara quanto ao calendário para a sua eliminação gradual.
- *Separação dos papéis da Sonangol de concessionária e de operadora.* Tanto o Governo como a Sonangol indicaram que não haverá alterações nesta configuração, pelo menos até 2010, com a justificação de que existem limitações institucionais e técnicas no Ministério das Finanças e no Ministério do Petróleo que impedem um progresso mais rápido.
- *Actualização da página do Ministério das Finanças na Web.* Em Março de 2006, foi prestada informação detalhada ao Banco e ao Fundo acerca das exportações de petróleo, preços do petróleo e lucro do petróleo pela Direcção dos Impostos no Ministério das Finanças relativamente ao ano de 2005, mas a informação na Internet continua desactualizada.
- *Envolvimento da sociedade civil nas questões de gestão das finanças públicas.* O MF pediu ao Banco Mundial que organizasse workshops de alto nível sobre gestão das receitas petrolíferas. Trata-se de uma iniciativa meritória, mas que precisa de ser complementada com mais acções perenes de envolvimento com a sociedade civil.
- *Endosso formal dos critérios EITI.* Angola tem sido muito cautelosa em anunciar a sua adesão formal aos princípios e objectivos da Iniciativa para a Transparência das Indústrias Extractivas (EITI), apesar do encorajamento nesse sentido do Banco, FMI e outros organismos bilaterais.

Um quadro onde constem as acções tomadas e as metas a serem cumpridas realça as acções adicionais necessárias. No Quadro E.1 descrevem-se as áreas onde se registou progresso e onde são necessárias medidas adicionais para se cumprir uma governação objectiva e os critérios de transparência, que são tidos como boas práticas.

Lidar com as Questões do Sector dos Diamantes

O êxito do desenvolvimento da indústria dos diamantes conduzido pelo sector privado em Angola requer algumas mudanças. Existem várias melhorias possíveis no sector dos diamantes, o qual, em grande medida, permanece rodeado de secretismo e de paternalismo. As áreas mais importantes onde são indispensáveis melhorias são: (i) um enquadramento legal, previsível e transparente, que defina adequadamente os direitos e as obrigações dos investidores; (ii) um pacote fiscal que seja competitivo e, ao mesmo tempo, equitativo para as partes envolvidas; (iii) garantia da posse de alvarás de mineração; (iv) reforço da capacidade do Governo para monitorizar e regular o sector; e (v) um compromisso firme com vista à liberalização da comercialização.

Quadro E.1. Um Quadro para se Apreciar a Governação e a Transparência no Sector Petrolífero

Critérios	O que foi feito até à data	Acções Adicionais Necessárias
Resolver o potencial para conflito de interesses (Sonangol como concessionária)	Sonangol está a circunscrever as actividades típicas de concessionária	Sujeito a uma capacidade institucional credível, transferir o papel de concessionário para o Ministério do Petróleo
Iniciar a supervisão governamental adequada da Sonangol	MF/MINPET têm pouca capacidade para fazerem essa supervisão	Contratar o apoio de consultor qualificado. Criar capacidade.
Verificar/incluir os fluxos financeiros da Sonangol no orçamento	Actividades de circunscrição e de auditoria de actividades parafiscais da Sonangol. As despesas próprias ainda não integram o orçamento. As despesas parafiscais obedecem aos procedimentos orçamentais, com um atraso de 90 dias	Incluir no orçamento as despesas próprias e parafiscais da Sonangol e observar os procedimentos orçamentais sem atraso.
Criar capacidade institucional adequada nos ministérios/organismos chave.	Com grave falta de recursos. Qualificações inadequadas. Salários baixos. Ministério das Finanças actualmente a utilizar um modelo de projecção de receitas petrolíferas, desenvolvido pela AUPEC.	Reforçar a capacidade técnica na DNI. Criar ou transferir capacidade para o MINPET. Resolver a questão dos ordenados.
Efectuar auditorias independentes e qualificadas dos pagamentos feitos e das receitas recebidas.	Custo anual da indústria e auditorias fiscais por auditores internacionais experientes. Auditor faz a verificação das declarações de impostos face às avaliações de impostos analisadas e aos pagamentos realizados e identifica as discrepâncias.	Tomar nota e actuar de acordo com as recomendações dos auditores. Assim, permitir-se-á uma comparação dos pagamentos feitos pela indústria e as receitas recebidas pelos governos federal e provincial. Auditar as receitas recebidas pelo MF, Cabinda e Zaire; incluir Sonangol e liquidar atrasados.
Publicação dos resultados da auditoria através de um meio de acesso fácil	Actualmente, publicação dos pagamentos detalhados da empresa na página do MF na WEB. Resultados da auditoria ainda não foram publicados.	Adicionar resultados da auditoria. Melhorar a acessibilidade ao sítio na WEB. Considerar um meio de comunicação de alcance mais vasto para a sua publicação.
O exercício de auditoria deverá aplicar-se a todas as companhias, incluindo a Sonangol	Prática corrente mas os dados da Sonangol provêm do operador do bloco.	Sonangol fornecerá dados directamente ao MF. Publicação das auditorias às empresas da Sonangol
Envolver a sociedade civil no processo de gestão e de transparência das receitas	Presentemente, não existe envolvimento	Workshops temáticos deverão incluir a sociedade civil. Estabelecer centros de informação pública independente

(Continued)

Quadro E.1. Um Quadro para se Apreciar a Governação e a Transparência no Sector Petrolífero (*Continued*)

Critérios	O que foi feito até à data	Acções Adicionais Necessárias
Introduzir clareza no enquadramento/procedimentos legais, contratuais e fiscais	Minutas e textos legais de acesso difícil. DNI está a preparar um Manual dos Impostos	Compilar e publicar textos e procedimentos legais. Manual dos Impostos.
Desenvolver um plano de acção calendarizado e financiado, para a implementação da agenda relativa à transparência	Não existe nenhum plano actualmente, embora tenham sido programadas componentes individuais.	Preparar e publicar um plano explícito

Independentemente da estratégia seleccionada pelas autoridades para promover o sector dos diamantes, deveria sempre enveredar-se por uma abordagem integrante. A médio prazo, as intervenções com vista a melhorar o subsector da mineração artesanal e de pequena escala deveriam centrar-se em: a) melhorar a qualidade de vida dos que vivem e trabalham nos campos de diamantes; b) reabilitar o ambiente e aumentar a produtividade agrícola; c) beneficiar as infra-estruturas regionais; d) aumentar a produtividade e a segurança na mineração; e e) incentivar um crescimento económico generalizado via efeito multiplicador da actividade de mineração e das melhorias nas infra-estruturas. Deveria também dar-se uma atenção especial a: (i) implementação continuada do Processo Kimberley; e (ii) harmonização das políticas internas com as dos países seus competidores e vizinhos (especificamente a República Democrática do Congo) para se desencorajar o contrabando e estimular o comércio através dos canais oficiais.

Como Revitalizar o Sector Privado

Há um universo de oportunidades de investimento em Angola tanto na economia mineral como na não mineral que se podem concretizar através da eliminação de barreiras e de constrangimentos ao clima de investimento. Com o advento da paz em 2002 o país está agora a ter que enfrentar o desafio de ter de canalizar os seus imensos recursos naturais para a reconstrução das suas infra-estruturas e actividades de redução da pobreza. No entanto, as melhorias na competitividade da indústria local e nos esforços para diversificar a economia permanecem prejudicadas pela infra-estrutura inadequada, fracos indicadores de governação e um ambiente de negócios menos do que apropriado. Para se resolverem estes problemas existem 5 áreas que vão precisar atenção:

- *Estruturar um diálogo entre os sectores público e privado.* Num prazo muito curto, a criação de um mecanismo de consulta entre o governo e o sector privado é fundamental para se assegurar uma agenda de crescimento liderado pelo sector privado.

- *Reduzir o número de regulamentação e procedimentos desnecessários.* Apesar de, nos últimos tempos, se ter reduzido a burocratização, mais terá que ser feito neste domínio. Uma boa maneira de começar uma segunda fase de reformas eficazes seria com a execução sistemática de um inventário das leis e regulamentos existentes que afectam o clima de investimento do país.
- *Atenuar as restrições relativas ao registo de propriedade e de imóveis.* É essencial que se facilite o registo de propriedade e de imóveis. As autoridades podiam facilitar o registo formal de imóveis com a redução do tempo e do custo requeridos para o seu registo.
- *Rever as leis do trabalho com vista a dinamizar o mercado de trabalho.* Em Angola, nos tempos que correm, quando um empregador decide, por razões operacionais, que precisa de despedir um empregado, precisa de um alto grau de persistência e de paciência. É um processo caro, moroso e trabalhoso. Contudo, é importante reduzir o custo dos despedimentos e, simultaneamente, assegurar a protecção dos direitos dos trabalhadores.
- *Manter o compromisso do governo de melhorar a governação e adoptar medidas de combate à corrupção.* O Banco Mundial calcula que um país, que melhore a sua governação de um nível relativamente baixo para um nível médio, a longo prazo poderá quase que triplicar o rendimento per capita da sua população e reduzir igualmente a mortalidade infantil e o analfabetismo. Quanto à corrupção, a pesquisa feita sugere que equivale a um imposto pesado que aumenta os custos e reduz a eficiência.

Opções para Reanimar a Agricultura

Avaliar os custos de produção com vista a reduzi-los. Na presença de uma taxa de câmbio real sobrevalorizada, o país deveria explorar as possibilidades de reduzir os custos unitários agrícolas. O país goza de melhor pluviosidade do que muitos dos seus vizinhos, tem rendimentos substancialmente menores e tem estado à margem do progresso tecnológico, há algumas décadas, no que toca a novas variedades ou outras áreas. A questão chave é, porém, se os lucros na produtividade podem ser suficientemente amplos para contrabalançar as desvantagens decorrentes de uma divisa forte e dos altos custos dos transportes. Como não há uma indicação clara quanto à estrutura do custo real da produção em Angola, só poderá responder-se a esta questão através de uma análise rigorosa, caso a caso, do orçamento de cada cultivo individual.

Aumentar a importância da agricultura no orçamento. Elevar a prioridade atribuída à agricultura e ao desenvolvimento rural através da afectação orçamental será um passo importante para o reforço do sector. As dotações orçamentais atribuídas ao Ministério da Agricultura e Desenvolvimento Rural (MINADER) em 2004 atingiram Kz2 300 milhões (aproximadamente US$25 milhões). Este valor corresponde a apenas 0,64% do total do orçamento nacional.

Avançar progressivamente no sentido de uma maior descentralização. Deveria estimular-se a descentralização administrativa uma vez que a agricultura talvez seja o exemplo de um protótipo de actividade, cuja melhor administração <u>não</u> se faz a partir da capital. O avanço no sentido da descentralização das despesas deveria, no entanto, realizar-se gradualmente para se evitar o desperdício e a falta de eficiência. São igualmente importantes as preocupações no que toca à capacidade das estruturas do governo local para estabelecer e operar os mecanismos administrativos e orçamentais necessários que a descentralização exige.

Desenvolver e apoiar um ambiente voltado para o mercado. Deveria também haver uma alteração das políticas no sentido de se aprofundar a desregulamentação do mercado. A curto prazo, implicaria um ajustamento e/ou eliminação dos programas de ajuda alimentar para se evitar a concorrência desleal com a emergente produção interna de cereais. Na área de comercialização e distribuição o governo ainda define as margens grossistas e retalhistas. Os controlos exercidos pelo Governo demonstraram, em Angola e noutros países, que se traduzem numa rede de comercialização pouco desenvolvida e em agricultores mal servidos.

Aumentar a eficácia do MINADER. Para poder funcionar devidamente e contribuir eficazmente para o desenvolvimento do sector da agricultura, o MINADER terá de sofrer mudanças destinadas a aumentar o desempenho. Para aumentar a sua eficácia, o Ministério beneficiaria com medidas destinadas a: (i) identificar posições redundantes e improdutivas para as eliminar; (ii) reformar a sua estrutura para reforçar os seus serviços externos e de pesquisa; (iii) criar uma unidade de alto nível de análise de políticas; e (iv) reforçar a actual unidade estatística do Ministério.

Criar incentivos mais atraentes com vista a estimular a competitividade. O desenvolvimento do sector exigirá investimentos em infra-estruturas e um melhor clima regulador. Uma visão realista do futuro da agricultura de Angola deverá incluir uma combinação de explorações agrícolas comerciais e familiares. Ambas precisam de melhores infra-estruturas de transportes, de um sistema de comercialização moderno e de um ambiente regulador propício.

Aumentar o Bem-Estar através da Prestação de Serviços

Houve um número de programas sociais oficiais que tiveram algum sucesso mas é preciso fazer mais para se chegar às áreas mais remotas e mais necessitadas do país. Conquanto os programas nacionais existentes administrados pelo Governo e pelo sector privado indiquem que se estão a cumprir os objectivos previstos, a maior parte deles não tem capacidade para responder plenamente às necessidades dos pobres pois a procura em muito excede a oferta. Para aumentar a eficácia dos programas sociais, actuais e futuros, as autoridades deveriam: (i) adoptar mecanismos de consulta aos pobres quanto ao modo de afectar fundos públicos; e (ii) utilizar a poupança fiscal para melhorar a prestação de serviços. Discutem-se abaixo estas duas opções.

Levar os Serviços Sociais até aos Pobres

Pode ser muito útil a existência de mecanismos de consulta aos pobres sobre como e onde se afectarem os fundos públicos. A experiência de Angola com o FAS é um exemplo excelente deste modo de funcionamento. Deveriam adoptar-se mecanismos de consulta automática aos pobres sobre a tomada de decisões relativa ao investimento público, para além dos canais de responsabilização com vista a assegurar o controlo da qualidade e a participação da sociedade civil.

Utilização das Poupanças Fiscais para se Melhorar a Prestação de Serviços

Os subsídios aos preços do petróleo e dos serviços de utilidade pública deveriam ser progressivamente eliminados e as poupanças reorientadas com vista a melhorar a qualidade dos serviços públicos e a compensar os pobres. Numa primeira fase, as autoridades precisam de anunciar um programa abrangente para lidar com os subsídios de uma forma progressiva, a partir de 2006. Numa segunda fase, o actual mecanismo de fixação de preços da Sonangol deveria ser reavaliado e feita a eliminação gradual do nível ajustado dos subsídios, enquanto se continuar a implementar o programa de compensação concebido durante a primeira fase. O programa deverá ser ajustado, em qualquer momento, no caso de haver uma queda dos preços internacionais do petróleo.

As autoridades não deveriam negligenciar os custos políticos das medidas. A evidência internacional demonstra que os aumentos dos preços do petróleo que não se façam acompanhar de um conjunto de medidas paliativas conduzem, tipicamente, a protestos públicos e podem desencadear violências e distúrbios sociais. A experiência de outros países demonstra que qualquer programa destinado a reafectar as despesas públicas deveria prestar uma atenção muito particular aos seguintes princípios de sentido amplo:

- *Atracção política:* Os esquemas de compensações, programas de despesas politicamente atraentes, são essenciais para a sustentabilidade política e social da eliminação de um subsídio. O anúncio das medidas de compensação, feito de uma forma credível e com uma grande visibilidade, é um aspecto indispensável do pacote.
- *Em prol dos pobres e eficazes:* O ideal seria que se tratasse de medidas que maximizam o impacto do desenvolvimento e da redução da pobreza. Os beneficiários dos programas de "compensação" deveriam ser os pobres e, possivelmente, outras pessoas que também tenham perdido bastante com a eliminação dos benefícios. É necessário que estes últimos também sejam contemplados.
- *Velocidade das despesas e impacto nas famílias:* A velocidade com que se concebem os programas e se gasta o dinheiro é importante por razões macroeconómicas, enquanto a velocidade com que os programas causam impacto nas famílias pobres é importante por razões de compensação e de economia política.

Uma das medidas paliativas que pode acarretar grandes benefícios para os pobres é uma política de água e saneamento. A análise de bem-estar efectuada no estudo patrocinado pelo Banco Mundial - Reino Unido sobre os impactos da eliminação gradual de subsídios indica que as alterações do preço da água produzirão um impacto social marginal muito superior ao das alterações dos preços dos combustíveis. Dada a situação precária das infra-estruturas, gestão e fornecimento do serviço, parece haver uma vasta margem para reforma no sector da água pública, com o intuito de o tornar mais eficiente, expandir o abastecimento de água à população e melhorar a qualidade da água. Os fundos necessários para financiar um empreendimento desta dimensão podiam provir dos fundos economizados com a eliminação gradual dos subsídios aos combustíveis e da participação do sector privado.

Enunciam-se abaixo outras políticas paliativas que podem ser implementadas:

- *Política de Educação Pública:* Promoção do sistema de vales, transferências de dinheiro condicionais e refeições escolares com vista a oferecer incentivos aos pais para mandarem os filhos para a escola e incentivos às crianças para que permaneçam na escola;

- *Política de Prestação de Cuidados de Saúde:* Promover o investimento em infra-estruturas, oferecer cuidados preventivos e clínicas móveis para aumentar o acesso aos serviços de saúde;
- *Política de Transportes Públicos:* Criação de um passe social que possa ser utilizado quer em autocarros quer em vans, concessão de subsídios ao serviço dos vans com base nas taxas de utilização, implementação de passes patrocinados pela entidade empregadora.

Um pacote para se eliminar gradualmente, com êxito, os subsídios em Angola, irá exigir compromisso político e uma boa estratégia de implementação e de monitorização. Com vista a implementar o pacote com sucesso, o Governo de Angola vai precisar de formar equipas e de lhes atribuir responsabilidades. As equipas deverão cobrir as áreas principais seguintes: (i) questões macroeconómicas e fiscais; (ii) programação das despesas (despesas compensatórias e com a protecção social e outras despesas com o desenvolvimento e a redução da pobreza); e (iii) socialização das poupanças.

Por último, para se saber se uma estratégia de redução da pobreza é eficaz no que toca à redução da pobreza, é necessário criar um sistema de monitorização da pobreza para se observar os indicadores chave, no tempo e no espaço. Em Angola há uma grande ausência de conhecimento sobre o modo de vida dos pobres. De uma maneira geral é difícil ter uma avaliação global dos programas existentes das redes de segurança uma vez que não têm, à partida, um número previsto de beneficiários. Esta falha advém da falta de fiabilidade dos dados relativos à pobreza, os quais deveriam servir de referência para os programas, com vista a avaliar a adequação e a eficácia no que respeita aos seus alvos e ao impacto das suas actividades. Uma forma razoável de se ultrapassar esta limitação seria com o reforço do Instituto Nacional de Estatística (INE) e a criação de um programa de monitorização anual da pobreza e de indicadores e divulgá-los ao público em geral.

Resumo das recomendações e passos seguintes

O Quadro abaixo descreve as áreas de acção prioritárias e as medidas seguintes que se sugerem (ver Quadro E.2). Algumas das recomendações já estão em fase de implementação, em muitos casos com o apoio do Banco Mundial.

Quadro E.2. Sumário dos Diagnósticos e das Recomendações

Áreas de Reforma	Passos Seguintes Recomendados
1. Reforçar o Desempenho Macroeconómico: O principal objectivo é reforçar as políticas económicas para se garantir uma estabilização macroeconómica sustentável e proteger a economia da volatilidade associada com os preços internacionais do petróleo (Capítulos 1, 2, e 3).	
Resolver as deficiências contínuas na concepção e implementação da política macroeconómica.	Adoptar um envelope fiscal prudente para os anos que se seguem e evitar aumentos rápidos das despesas públicas. Continuar a adoptar, no orçamento, um preço do petróleo conservador.
	Preparar um MTFF. Calcular tectos anuais para as despesas públicas com base nas projecções de receitas a médio prazo, calculadas em função do preço do petróleo a longo prazo. Deslocar-se gradualmente no sentido de um MTEF.
	Criar regras fiscais claras para a recém anunciada conta de reserva do petróleo, com base em princípios sólidos de governação e de transparência. Adoptar previamente um valor máximo que sirva de limite ao volume de recursos a ser acumulado.
	Evitar a tendência pró-cíclica dos gastos fiscais. Gerar excedentes fiscais quando os preços do petróleo estiverem a subir para se poder sustentar as despesas quando as receitas eventualmente baixarem.
	Utilizar as receitas petrolíferas em investimentos produtivos e não no consumo (aumentar a capacidade produtiva) para se reduzirem os custos da totalidade da economia.
	Adoptar uma âncora monetária clara, utilizar as vendas de divisas estrangeiras para reduzir a liquidez e manter a pressão descendente na inflação e não oferecer resistência à eventual apreciação da moeda.
	Constituir reservas em divisas e reduzir a dívida externa.
	Modernizar as políticas e instituições na área de aquisições, de gestão financeira e de avaliação de programas.
2. Solucionar Questões Pendentes no Domínio da Governação e Transparência: Uma estratégia sólida para a gestão da riqueza mineral do país exige transparência e uma governação salutar nos sectores do petróleo e dos diamantes para se evitar a procura de ganhos ilícitos e o desperdício (Capítulos 3 e 4).	
Lidar com conflitos de interesse no seio da Sonangol e da Endiama.	Prosseguir uma fiscalização adequada da Sonangol e da Endiama, por parte do Governo.
	Separar as actividades da Sonangol e da Endiama como operadoras e como reguladoras nos seus respectivos sectores, mediante a transferência das responsabilidades reguladoras para os ministérios apropriados.
Melhorar a governação e a transparência nos sectores do petróleo e dos diamantes.	Apoiar a unidade de gestão das receitas petrolíferas dentro do Ministério das Finanças e preparar os funcionários para trabalharem com o modelo de projecção das receitas petrolíferas.
	Publicar as auditorias do grupo Sonangol e Endiama.
	Definir um plano e um calendário para a eliminação integral das operações parafiscais da Sonangol.
	Actualizar imediatamente as informações relativas às exportações de petróleo, preços do petróleo e lucro do petróleo na página do MINFIN na Internet.

(Continued)

Quadro E.2. Sumário dos Diagnósticos e das Recomendações (*Continued*)

Áreas de Reforma	Passos Seguintes Recomendados
	Endossar formalmente os critérios EITI.
	Compilar e publicar textos e procedimentos legais e bem assim o manual de tributação relativo ao enquadramento fiscal/legal do sector petrolífero. Promulgar legislação transparente e estável no âmbito de uma nova lei para o sector dos diamantes.
	Encomendar a actualização do Estudo de Diagnóstico dos Diamantes de 2003.
	Rever e regularizar os direitos minerais existentes (exploração e mineração).
Melhorar o clima de negócios no sector dos diamantes.	Produzir melhores informações geológicas e disseminá-las
	Liberalizar a venda de diamantes; identificar e avaliar opções para a comercialização de diamantes no mercado livre; fazer testes piloto de opções seleccionadas para a transacção de diamantes num mercado livre.
	Sensibilizar e divulgar as regras do jogo, tanto local como internacionalmente.
	Reforçar a implementação do Esquema de Emissão de Certificados de Origem.

3. *Promover o Desenvolvimento do Sector Privado:* *O clima de negócios em Angola é tido como um dos menos propícios do mundo e, qualquer estratégia para apoiar o crescimento do sector privado irá precisar de promover reformas que sejam favoráveis à criação de um ambiente de negócios atraente (Capítulo 5).*

Eliminar barreiras e constrangimentos ao clima de negócios.	Estruturar um diálogo entre os sectores público e privado.
	Reduzir o número de regulamentos e procedimentos desnecessários.
	Moderar as restrições à propriedade e registo de imóveis e reduzir os respectivos custos.
	Rever as leis do trabalho e dinamizar o mercado de trabalho.
	Manter o compromisso do governo de melhorar a governação e de adoptar medidas anti-corrupção.
Utilizar o comércio como um padrão para se medir o sucesso na diversificação.	Resistir às pressões proteccionistas que vão surgir com a apreciação da taxa de câmbio.
	Desenvolver sistemas de incentivos à exportação onde existirem vantagens comparativas.
	Aumentar a facilitação do comércio e reforçar a capacidade do comércio.
	Tirar partido de acordos de comércio e participar activamente em negociações regionais e globais.
Introduzir esquemas mutuamente vantajosos para promover a interacção entre as empresas petrolíferas e os fornecedores locais.	Preparar estratégias de desenvolvimento sectorial.
	Investir na formação profissional dos trabalhadores angolanos.
	Promover *joint ventures*.
	Melhorar o clima de negócios, a competitividade e os mecanismos de escoamento.

(*Continued*)

Quadro E.2. Sumário dos Diagnósticos e das Recomendações (*Continued*)

Áreas de Reforma	Passos Seguintes Recomendados
4. *Reforçar o Sector Agrícola: A agricultura pode desempenhar um papel importante no futuro de Angola mas esse potencial só poderá concretizar-se através das tão necessárias reformas institucionais e de uma maior prioridade ao sector no orçamento do estado (Capítulo 6).*	
Aumentar a competitividade na agricultura.	Analisar as políticas de apoio à agricultura (comercial e fiscal) com vista à formulação de uma estratégia clara para aumentar a competitividade do sector.
	Apoiar os pequenos agricultores e resistir à tentação de se envolver em actividades públicas de produção ou de comercialização.
	Reparar estradas nas zonas rurais para facilitar o acesso aos mercados.
Eliminar os obstáculos ao crescimento da produção	Preparar uma análise detalhada das estruturas de custo actuais por cada cultivo com vista a reduzir os custos de produção respectivos.
	Aumentar a importância dada à agricultura no orçamento.
	Deslocar-se progressivamente no sentido da descentralização administrativa e das despesas.
	Desenvolver e apoiar um ambiente do sector dirigido para o mercado; rever os controlos de comercialização existentes; eliminar gradualmente o envolvimento do governo nas actividades de produção e de comercialização; rever a política actual de distribuição e de subsídios à venda de fertilizantes.
Aumentar a eficácia do MINADER	Efectuar um estudo global da força de trabalho do MINADER para se identificar as posições redundantes ou improdutivas e formular um plano para a sua eliminação.
	Reformar a estrutura do MINADER com vista a reforçar os seus serviços externos e de pesquisa.
	Criar uma unidade de alto nível para a análise de políticas dentro do Ministério com capacidade para analisar políticas alternativas e avaliar projectos.
	Melhorar a actual unidade estatística do Ministério que deverá servir de elemento de ligação com o Instituto Nacional de Estatística e com o Ministério do Plano para conceber um levantamento agrícola nacional.
5. *Melhorar a Prestação dos Serviços Públicos: Aumentar os níveis de vida através das despesas públicas e da prestação de serviços sociais e económicos (Capítulo 7).*	
Elaborar um programa gradual de obras públicas.	O objectivo é um programa de emprego nas obras públicas para a mão-de-obra pouco qualificada.
Aumentar o número de oportunidades de emprego para a força de trabalho.	Rever a legislação laboral para se introduzir dinamismo no mercado de trabalho (mas com respeito pelos direitos dos trabalhadores).
	Apoiar a organização de pequenas empresas em cooperativas.
Encorajar uma maior participação comunitária na concepção e implementação de investimentos no domínio de serviços sociais.	Adoptar mecanismos de melhores práticas para a consulta aos pobres no que toca à tomada de decisões sobre investimento público.
	Aumentar o âmbito e a cobertura do FAS.
	Promover um maior grau de coordenação entre os parceiros de desenvolvimento e o governo (local e provincial).

(*Continued*)

Quadro E.2. Sumário dos Diagnósticos e das Recomendações (*Continued*)

Áreas de Reforma	Passos Seguintes Recomendados
Utilizar as poupanças conseguidas com a eliminação gradual de subsídios para se melhorar a prestação de serviços.	Preparar um programa para a eliminação progressiva dos subsídios aos preços dos combustíveis e aos preços dos serviços de utilidade pública. Adoptar uma estratégia em duas frentes com vista a: (i) colocar a factura dos subsídios aos preços dos combustíveis em linha com o preço implícito do petróleo no orçamento de 2006: e (ii) analisar o mecanismo de fixação de preços da Sonangol para se eliminar por completo os subsídios aos combustíveis e liberalizar a distribuição.
	Utilizar a poupança fiscal para compensar os pobres e todos os que perderam com a eliminação dos subsídios mediante o investimento na reabilitação de infra-estruturas no sector da água; educação pública; prestação de cuidados de saúde; e transportes públicos.
	Criar uma equipa interministerial para preparar o pacote. Ao preparar o pacote, a equipa deveria ter em conta as áreas seguintes: questões macroeconómicas e fiscais; programação de despesas (gastos com a protecção social e de carácter compensatório); e socialização da poupança.
Estabelecer um programa de monitorizaçao anual da pobreza e indicadores sociais.	Reforçar o Instituto Nacional de Estatística (INE) com melhor capital físico e humano. Poderão ser necessários salários mais altos para atrair pessoal qualificado.
	Rever o plano nacional de estatística e dar maior prioridade à monitorização da pobreza.
	Uma vez em vigor um apoio adequado, conduzir e disseminar levantamentos anuais da força de trabalho.
	Enquadrar os planos de despesas públicas numa estratégia a médio prazo para redução da pobreza e para prossecução das Metas de Desenvolvimento do Milénio (MDG).

Executive Summary

The Report's Main Messages

This *Country Economic Memorandum* identifies six core areas where a strategic approach for the development of a broad-based growth strategy is required. In designing a sustainable development strategy for the country, the authorities will need to address a number of issues in six major areas which have been found to constrain growth and equitable development in Angola: (i) the incomplete transition to a market economy; (ii) macroeconomic management; (iii) governance and transparency in the management of the mineral wealth; (iv) the business environment; (v) agriculture; and (vi) public service delivery to the poor. The recommended strategic approach to each one of these areas is summarized below. The subsequent sections of this executive summary highlight the broader conclusions and the main recommendations of the *Country Economic Memorandum*.

- First, Angola needs to complete the transition to a market economy. With the advent of Independence in 1975, the Angolan government opted for a centralized economic system which today is believed to have contributed, together with the growing dependence on oil, to constrain the development of sound institutions that foster the appearance of a vibrant private sector. The transition to a full market economy system will require political commitment at the highest levels since entrenched vested interests will have to be dismantled. Additionally, the country will also need to complete the transition to a multi-party democracy which will help to accommodate social and ethnic tensions that permeated and fueled civil conflict in Angola in the past and which remain latent in the present. These points are treated in Chapters 1 and 2.
- Second, on the macroeconomic front, continuing deficiencies in policy design and implementation, especially at the aggregate level, still need to be addressed. Recent progress in the management of the economy is encouraging, most notably the success in reducing inflation and strengthening the government's fiscal position. However, these have been largely influenced by rather favorable developments in the oil sector. To guarantee the sustainability of the recently achieved macroeconomic stabilization Angola needs to put in place a sound economic policy mix with a focus on better public expenditure management and administer some tensions that are common to economies that are transitioning from war to peace and from a command setting to a market economy. Chapters 2 and 3 present advice on macroeconomic policy options.
- Third, a clearer strategy to manage the country's growing oil and diamond wealth based on sound governance and transparency principles must be defined. The Angolan economy will experience a massive oil revenue windfall with a concomitant fiscal gain over the next two decades or so. Diamond production is also projected to grow strongly. The derived gains from increased production, however, represent the depletion of the country's oil and diamond reserves. This raises the question of how to use the incomes from non-renewable resources to create income sources into the future, including for future generations. Sound governance and

transparency principles should guide the establishment of any natural resource management strategy in Angola. Issues and options on this area are discussed in detail in Chapters 3 and 4.

- Fourth, and related to the point above, there is a real urgency to improve the business environment and the investment climate in Angola. The business environment in the country is one of the least favorable in the world. If the authorities want to promote a broad-based economic recovery of the country, with more jobs and higher incomes for the average Angolan, then pro-business measures that can enable companies to compete more effectively in an open economy need to be implemented fast and social programs targeting the most vulnerable population and insulating them from adjustment costs should be put in place. Otherwise, the Angolan economy will continue to be highly dependent on mineral resource exploitation and the majority of the population will continue untouched by the mineral wealth. Chapter 5 focuses on the challenges and opportunities associated with the development of the non-mineral private sector in Angola.
- Fifth, considering areas of potential outside the mineral sectors, the importance of agriculture as a source of employment and incomes should not be neglected. Angola is a natural food and cash crop producer and agriculture holds the potential to contribute for further economic growth in the years to come. Government policies should support the development of smallholders but, at the same time, foster a conducive environment to encourage investments in the private commercial sector. Chapter 6 outlines a strategy to increase agriculture output and improve the incentives that can contribute to higher competitiveness in the sector.
- Finally, as part of the peace dividend to the Angolan population, the quality and supply of public services, especially to the poor, must improve. As a post-conflict country, Angola faces a huge challenge to improve the welfare of its population, including the poorest. The country stands a very good chance of meeting this challenge owing to the increasing resources available from the exploitation of its natural resources. Feasible options to improve the welfare of the majority of the Angolan people are discussed in Chapter 7.

The Road Ahead Involves Risks

The road ahead involves political risks which will be difficult to address. Ample empirical evidence has shown that natural resource dependence is particularly problematic, since it is easily captured by the ruling elite, removing the incentive for the government to actively engage its citizenry. This destroys both the capacity and the legitimacy of the state, exacerbating social divisions and even leading to direct conflict over the resource itself. Research at the World Bank and elsewhere points to resource dependence as one of the most important causes of civil wars (see Figure E.1).[3] In the case of Angola, moving forward with the reform agenda in the areas of governance, transparency, public finance management, business

---

3. Violent secessionist movements can often be traced to oil. Examples include Aceh in Indonesia, Biafra in Nigeria, conflicts in Sudan, Chad, Congo Brazzaville, and in Angola itself (see Busby et al., 2002).

environment, and public service delivery will be politically challenging, but the risks of not reforming at all or wait too long to take action can be devastating given the huge gaps that the country needs to fill. Notwithstanding, while an increase in natural resource revenues can potentially increase rent-seeking behavior and give governments an excuse to delay reform, it can also allow reform-minded governments to implement changes. Angola has now a real opportunity to use its natural resource wealth to mitigate tensions and adopt the difficult measures that can put the economy on a path of sustainable development. The World Bank stands ready to support the government in that endeavor.

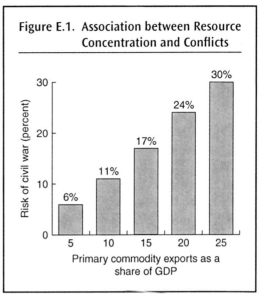

Figure E.1. Association between Resource Concentration and Conflicts

Source: Bannon and Collier (2003).

Socio-Economic Realities

A World Class Oil and Diamond Producer

Once one of the world's largest staple food producers, Angola is now known as a major oil exporter. Prior to independence, Angola was better known as a coffee exporter than as an oil exporter. Angola was not only the world's fourth largest coffee exporter, but also exported over 400,000 MT of maize annually, making it one of the largest staple food exporters in Sub-Saharan Africa. The structure of the economy then changed substantially after 1973 as the mining and service sectors increased their share in the composition of GDP. The country is now the second largest oil producer in Africa.

Angola is also the world's fourth largest producer of rough diamonds in terms of value, with the potential to become one of the leading global diamond producers. The country has close to 12 percent of the share of the world market and a high proportion of its production is of gem quality. Diamond reserves were estimated in 2000 at 40 million carats in alluvial deposits, and 50 million carats in kimberlite pipes, which are just now beginning to be exploited. Around 700 kimberlites of varying sizes (10–190 hectares) and shapes are known in Angola, aligned along a SW to NE trend across the country and into the Democratic Republic of the Congo.

In addition to oil and diamonds, the country is well endowed with agricultural resources which remain mostly untapped. Staple crops range from cassava in the humid north and northeast to *maize* in the central highlands and sorghum/millet in the dryer southern provinces. *Potatoes* are an important crop in the central plateau and *rice* is also grown over large areas in the north. *Cattle* is raised over broad areas in the central plateau but are particularly important in the southern provinces of Cunene, Huila, and Namibe where there are an estimated 3 million heads of cattle. *Coffee,* the most important cash crop during colonial times, grows well in the highlands from Uige and Malange through Kuanza Norte and as far south as Huambo and Bie. It is also important to note that some two thirds of the Angolan population live in rural areas and earn their living from agriculture, which currently accounts for less than 10 percent of GDP and receives less than 1 percent of all budget outlays.

Daunting Poverty and Welfare Indicators

Despite the country's significant natural wealth, existing social indicators still reflect low living standards. According to both the 2001 IDR (Income and Expenditures Survey) and the 2002 MICS (Multi Indicator Cluster Survey), approximately 70 percent of the population lives on less than 2 dollars a day and the majority of the Angolans lack access to basic healthcare. About one in four Angolan children die before their fifth birthday, 90 percent of whom perish due to malaria, diarrhea or respiratory tract infections, the maternal mortality rate (at 1,800 per 100,000 births) is one of the highest in SSA, and three in five people do not have access to safe water or sanitation. The HIV/AIDS prevalence rate is, according to official statistics, relatively low, affecting an estimated 3.9 percent of adults.[4] In terms of education, primary school enrollment is very low at 56 percent, and suffers from late entries into school and high repetition and drop out rates. Some 33 percent of the population is currently illiterate, though in rural areas this climbs to as many as 50 percent.

A Challenging Business Environment

The business environment in Angola is challenging. According to the 2006 *Doing Business* survey of the World Bank, establishing a company in Angola takes an average of 146 days, more than twice the regional average. Licensing is a time-consuming and costly procedure. The time to comply with all licensing and permit requirements is estimated at 326 days — almost one year. Registering property also takes about 11 months and costs more than 11 percent of the property value. Investor protection is not high when compared to similar countries and obtaining credit is equally difficult. The average import requires 10 documents, 28 signatures, and 64 days. It is equally difficult to enforce a contract—in calendar days, it takes 1,011 days to resolve a dispute starting from the moment a plaintiff files a lawsuit in court until settlement or payment. Dispute resolution among Angola's neighbors and potential competitors takes much less time—in Zambia 274 days; in Namibia 270; and in Botswana 154.

Besides having a difficult business environment, Angola ranks amongst countries with the worst corruption perception indices. In the rankings established by Transparency International's widely referenced Corruption Perceptions Index, petroleum exporting developing countries find themselves in the bottom one third of the countries listed. Angola's rank in the most recent release of the CPI was 151, number nine from the bottom. This is a particularly serious finding because corruption is recognized as one of the largest single inhibiting factors to private sector investment and growth.

---

4. Low estimates are 1.9% and high estimates are 9.4% (UNAIDS 2004).

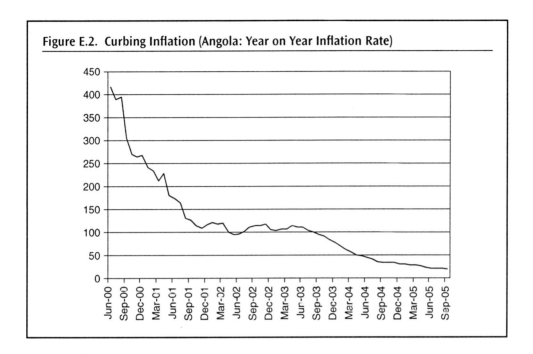

Figure E.2. Curbing Inflation (Angola: Year on Year Inflation Rate)

Macroeconomic Reform Agenda

Tangible Progress and Promising Opportunities

More recently, government's efforts to reduce inflation have been successful. Between 1999 and the peace agreement of 2002, annual consumer price inflation fell from around 300 percent to around 100 percent (see Figure E.2). Following the adoption of a stabilization program in September 2003, inflation fell sharply again and by December 2004 the 12-month inflation rate had declined to 31 percent. The improvement was largely due to the government's avoidance of money creation for deficit finance together with smaller fiscal deficits. In 2005, the cumulative rate of inflation dropped to 18.5 percent and the projection for 2006 is of an annual rate of 10 percent.

Despite the recent progress, there should be a stronger emphasis on the continuing deficiencies in policy design and implementation. The stabilization obtained so far needs to be strengthened with improved coordination of the fiscal policy with monetary and exchange rate policies. These policies need to spell out a consistent strategy to absorb the upcoming oil windfall without inhibiting growth outside the mineral sectors. A first step in the right direction would be to reduce public expenditures on consumption and increase spending on productive investments (e.g., infrastructure) of the sort that will have the effect of lowering domestic costs for the entire economy. To a certain extent, this strategy is already being followed by the authorities, but there are areas in which complementary measures are required and these are discussed below.

Aligning Economic Policy with Macroeconomic Realities

Instruments to smooth out the impact of the oil cycles should be adopted. The combination of fiscal, financial, trade, and monetary policies must care about stimulating private investments to increase the size and sophistication of non-extractive sectors. This will

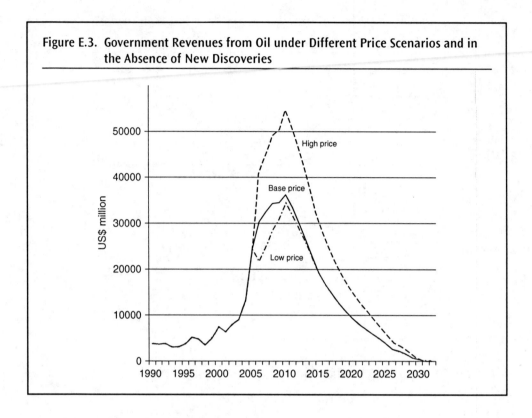

Figure E.3. Government Revenues from Oil under Different Price Scenarios and in the Absence of New Discoveries

require (i) prudence in the management of aggregate demand (i.e., avoid rapid expansion in public spending or credit, which can lead to macroeconomic imbalances); and (ii) the use of a medium-term economic framework to guide economic policy decisions. Instruments to smooth out the impact of the oil cycles in the economy and create a desirable pace of non-oil industrialization include:

- ■ *A prudent fiscal envelope for the coming years.* It is important to remember the hump-shaped profile of oil-related fiscal revenues and the fact that they will decline rapidly during the next decade, in the absence of new discoveries (see Figure E.3). Such a path would require running substantial surpluses when the rapid extraction of oil reserves takes place over the rest of this decade and most of the next decade. However, this may not be politically feasible. In addition, expectations about future growth in the non-oil sector may fail to materialize altogether, while the present discounted value (PDV) of the future oil wealth is uncertain and subject to oil price volatility. Therefore, although there may be ample room for public expenditure increases, the above-mentioned concerns call for a conservative fiscal envelope.
- ■ *A medium-term economic framework.* Currently, the horizon for decisions that affect macroeconomic policy is, at most, two years ahead as spending programs in annual budgets are determined only by revenue prospects for the coming year. Despite the fact that the government has been adopting a conservative approach towards specifying the oil price on which revenue projections are based, this approach can still yield sharp cycles in spending which can prove unsustainable in the long run. A preferred approach would be to base annual ceilings for public expenditure on medium-term

revenue prospects, evaluated at a long-run oil price. Under this approach, the recommendation of the previous paragraph of generating consistent fiscal surpluses could be accomplished over time so that financial reserves could be accumulated during the years of peak oil production to sustain spending when oil revenues eventually fall.

- *Anti-cyclical fiscal policies in relation to oil prices.* The pro-cyclicality of fiscal expenditures and oil prices is dangerous and can transmit volatility to the rest of the economy. A key policy objective for Angola should be to pursue fiscal strategies aimed at breaking the procyclical response of expenditures to volatile oil prices. In this sense, the approach currently used by the authorities of adopting a conservative oil price in the preparation of the budget is welcome, but with the remark that it should be complemented with the recommendations of the previous paragraphs.
- *Rapid and bold improvements in procurement practices.* In simple terms, the level of spending should be determined taking into account its likely quality and the capacity of the administration to execute it efficiently. In this regard, an abrupt enlargement of expenditure programs associated with oil windfalls carries important risks. A hasty public spending program may exceed the government's planning, implementation, and management capacity, with the result that it may be difficult to prevent wasteful spending.

Institutional Options to Manage the Windfall

An institutional framework will be required in order to effectively manage Angola's oil wealth. Regardless of whether or not Angola adopts the above recommendations, the Government must develop the institutional capacity required to manage future revenue. The recent announcement of the creation of an oil reserve account at the Central Bank that would accumulate part of the oil revenue windfall, and the use of an oil revenue forecasting model by the Ministry of Finance are welcome developments which need to be complemented by additional arrangements.

The establishment of rules for the oil reserve account should draw on the successes and failures of past experience in other oil producing countries. As the Government of Angola is considering the establishment of an oil reserve account, the following features, which are based on other models now being developed and implemented in other countries, might be considered useful on the way forward (a summary of these features in presented in Box E.1 below):

- The oil reserve account should be the designated recipient of petroleum windfall revenues. Partial capture of oil revenues by various ministries or state-owned entities would make it difficult to obtain efficient and sustainable management of the revenues.
- Transfers to the budget should be based on the agreed savings/expenditure rule. Without clear rules, transfers risk becoming increasingly dependent on short-term economic and political cycles, which would critically undermine the account's function.
- The BNA should be designated as the operational (day-to-day) manager for the account. An efficient, transparent, and rule-bound management requires dedicated and neutral staff, placed within the Central Bank.

> **Box E.1: Elements of a Revenue Management Framework for Angola**
>
> **Revenue consolidation and collection**
> – The collection of all petroleum related revenues are consolidated through the oil reserve account
> – Revenues are accounted for according to agreed and transparent accounting guidelines
> – Revenues are published in an accessible and timely manner
>
> **Define savings and consumption**
> – Transfers from the oil reserve account to the budget are based on the agreed savings/expenditure rules.
> – The rules are clear, predictable, and public, do not depend on administrative or political discretion for their application
> – In the development of guidelines for savings and consumption, macroeconomic and sustainability concerns are paramount
>
> **Institutionalize the transfer mechanism**
> – Transfers from the revenue collecting authority to the budget and the oil reserve account follow predefined rules and occur automatically, independent of administrative or political discretion
>
> **Account management**
> – Management of the funds is based on clear, transparent, agreed, and predictable rules, which allocate clear responsibilities and reporting requirements.
> – The BNA is designated as the operational (day-to-day) manager for the account.
> – The rules for investing the account's assets must be clear, agreed and published.
> – Fund assets should to a large extent be invested abroad and in safe instruments
> – The BNA reports on performance of the oil reserve account and asset allocation according to a preset schedule, and the reports are made public in an accessible way
> – A high level oversight committee (key ministries plus qualified external advisers) must be established.
> – The oil reserve account, its management, and guidelines must be subject to rigorous transparency requirements

- The rules for investing the account's assets (such as low-risk overseas treasury instruments) must be clear, agreed and published.
- A high level oversight committee (key ministries plus qualified external advisers) must be established.
- There must be annual, professional audits of the oil reserve account, and the auditors' report should be published.
- The oil reserve account, its management, and guidelines must be subject to rigorous transparency requirements.

Strengthening Governance and Transparency

Appropriate management of oil and diamond wealth in Angola will require improvements in the areas of governance, transparency and institutional capacity. In an economy heavily dependent on mineral resources, the presence of weak institutions may invite rent seeking and waste. In the case of Angola, where the level of institutional quality is low, in part because of the effects of the war and in part because of problems associated with the so-called Paradox of Plenty, it is extremely necessary to improve the factors that can affect the quality of governance. This will require further efforts in the following 4 broad areas.

Dealing with Conflicts of Interest

The relationship between Sonangol, the national oil company, and the Government creates conflicts of interest and is against good practice in the management of public finances. Sonangol performs multiple roles vis à vis the Government, including activities which would normally be performed by the Treasury and the Central Bank. It is a taxpayer, it carries out quasi-fiscal activities, it invests public funds, and, as concessionaire, it is a sector regulator. This multifarious work program creates conflicts of interest and characterizes a complex relationship between Sonangol and the government that weakens the formal budgetary process and creates uncertainty as regards the actual fiscal stance of the state.

A similar concern is associated with the operations of Endiama, the national diamond company. The Ministry of Geology and Mines is responsible for the implementation of the legal and regulatory framework for the sector, for issuing mineral rights, and for the geologic survey, while the mandate to approve kimberlite concessions is of the Council of Ministers. Endiama is by law the largest shareholder in all new diamond ventures and also has regulatory functions over the selection of companies that are to be granted new diamond mineral rights, the negotiation of mining contracts, and the monitoring and control of activities of diamond ventures. The company is, therefore, both an operator and a regulator in the diamond industry.

There is, thus, a need for institutional reforms in the oil and diamond sectors. In the oil sector, the role of Sonangol should be reassessed with a view to eliminate the conflict of interest and improve the quality and effectiveness of public finance management in Angola. In the diamond sector, it will be necessary to introduce a transparent licensing, regulatory, and tax structure that would avoid conflicts of interest and privileged access to development rights. These issues are discussed separately below.

Reassessing the Role of Sonangol

More needs to be done to phase out the dual role of Sonangol. Although there have been initiatives to add transparency in the relations between Sonangol and the Government, the current procedures have several disadvantages, in addition to that of recording fiscal activities outside the budget where they belong: a) as affirmed above, it is not transparent; b) it leads to frequent disputes; and c) it misrepresents Sonangol's position on taxes. The GoA is now moving to reflect quasi-fiscal activities in the budget, and Sonangol is taking steps in its external audit to identify and audit all such activities undertaken on behalf of the Government. Both of these measures are welcome. However, the offsetting mechanism is still very much a current practice and concrete efforts to phase it out have yet to materialize.

Sonangol can contribute more effectively to the development of the country by focusing on its core business. It is a recognized fact that the activities performed by Sonangol on behalf of the government create an additional administrative and operational burden on the company. If these were relinquished and Sonangol focused exclusively on its core business, the company could contribute even more to the development of the country. A few examples of areas where the increased intervention of Sonangol could make a difference include: (a) investments to increase fuel storage capabilities; (b) investments to improve fuel distribution to the interior of the country; (c) training of Angolan labor force to work in the IOCs; and (d) investments in social projects.

Addressing Governance and Transparency Issues

There has been some progress in improving governance and transparency in the petroleum sector since 2002. A number of recommendations made by the Bank in the past, both in the Oil Diagnostic Study and in the PEMFAR have been adopted and these are summarized below:

- The government has been publishing details of oil payments received (by block, by type of payment, with annual summaries by company) on the Ministry of Finance website, although with a significant lag of 6 months on average.
- The audits of the petroleum sector are being conducted. The financial statements of companies in the Sonangol group were audited comprehensively for the first time, by Ernst & Young, in 2003. The 2004 audit was recently completed. The auditors' reports, however, have yet to be published.
- Recent audits have followed acceptable accounting rules. The audits mentioned above applied Angola's recently issued accounting legislation where appropriate, and IAS rules where the new legislation does not apply. By end-2006 Sonangol expects to have moved fully to IAS standards.
- The roll out of the Angolan IFMS (SIGFE) is progressing steadily. Progress towards the implementation and full roll out of the Angolan IFMS (Integrated Financial Management System) has been steady and currently most of the expenditure side of the government accounts is registered in the system. However, progress in the inclusion of revenue data into the system remains very weak.
- The government is using a model for oil revenue forecasting. The Ministry of Finance has signed a contract with AUPEC, Aberdeen University Petroleum Economic Consultancy, to implement an oil revenue forecasting model and advise DNI on strengthening its capacity with respect to petroleum taxation.

But progress has been slow in addressing other equally important structural financial management issues. The Government has started to strengthen the capacity of the Ministry of Finance to control expenditures with the roll out of the SIGFE and the setting up of a fiscal programming unit. An initiative to gain some oversight over the operations of Sonangol on behalf of the Treasury by registering them ex-post in the SIGFE is also in place. But more needs to be done to arrive at a point of "normalization". Other suggested improvements in public financial management that remain to be addressed include:

- *Sonangol's quasi-fiscal operations:* Sonangol's quasi-fiscal operations are currently reflected in the budget (with a lag), but there is no clear indication about a timeframe for phasing them out.
- *Separation of concessionaire and operator roles of Sonangol.* The Government and Sonangol have both indicated that there will be no change on this configuration at

least until 2010 under the justification that there are institutional and technical limitations in the Ministry of Finance and in the Ministry of Petroleum that prevent faster progress.
- *Updating of the Ministry of Finance website.* Detailed information was provided to the Bank and Fund in March 2006 about oil exports, oil prices, and profit oil from the Tax Directorate of the Ministry of Finance for 2005, but the information on the website continues outdated.
- *Engagement with civil society on public finance management issues.* The MOF has asked the World Bank to organize high level workshops on petroleum revenue management. This is a welcome initiative that needs to be complemented with more perennial actions of engagement with civil society.
- *Formal endorsement of the EITI criteria.* Angola has been cautious about announcing formal adherence to the principles and objectives of the Extractive Industries Transparency Initiative (EITI), despite the encouragement of the Bank, IMF and bilaterals.

A scorecard of actions taken and targets to be met spotlights required further actions. Table E.1 describes areas where progress has been made and where further actions are required to meet objective governance and transparency criteria which are considered good practice.

Tackling Issues in the Diamond Sector

Successful development of the diamond sector led by the private sector in Angola will require change. There are a lot of improvements possible in the diamond sector that, to a large extent, remains secretive and victim of patronizing. The most important areas where improvements are called for include: (i) predictable and transparent legal framework that adequately defines investors' rights and obligations; (ii) a fiscal package that is competitive and at the same time equitable for the concerned stakeholders; (iii) security of tenure of mining permits; (iv) strengthened capacity of the Government to monitor and regulate the sector; and (v) a firm commitment towards marketing liberalization.

Whatever the strategy selected by the authorities to promote the diamond sector, an integrative approach should be pursued. Over the medium term, the interventions for improving the artisanal and small scale mining sub-sector should focus on: a) improving the quality of life of those living and working in diamond fields; b) rehabilitating the environment and improving agricultural productivity; c) upgrading regional infrastructure; d) improving mining productivity and safety; and e) encouraging general economic growth via multiplier effect of mining activity and infrastructure improvement. Additional emphasis should also be given to continued implementation of the Kimberley Process, and the harmonization of the domestic policies with those of competing and neighboring countries (namely the Democratic Republic of the Congo), to discourage smuggling and encourage trading through official channels.

How to Strengthen the Private Sector

There is a world of opportunities for investment in Angola both in the mineral and non-mineral economies that can be realized by removing barriers and constraints to the investment climate. With the advent of peace since 2002 the country now faces the daunting challenge of channeling its huge resource endowment into reconstruction of its infrastructure

Table E.1. A Scorecard to Assess Governance and Transparency in the Oil Sector

Criteria	What Has Been Done So Far	Required Further Action
Resolve conflict of interest potential (Sonangol as concessionaire)	Sonangol is ring-fencing concessionaire activities	Subject to credible institutional capacity, transfer concessionaire role to Ministry of Petroleum
Introduce adequate Government oversight of Sonangol	MOF/MOP have little capacity to provide oversight	Engage qualified consultant support. Build capacity.
Reconcile/include Sonangol financial flows with budget	Sonangol ring-fencing and auditing quasi-fiscal activities. Own expenditures are still outside budget. Quasi-fiscal expenditures comply with budget procedures with a 90-day lag/	Bring Sonangol quasi-fiscal and own expenditures into budget and comply with budget procedures without delays.
Create adequate institutional capacity in key ministries/agencies	Seriously under-resourced. Inadequate skills. Salaries low. Ministry of Finance now using oil revenue forecasting model developed by AUPEC.	Strengthen technical capacity in DNI. Build or transfer capacity to MOP. Resolve salary issue.
Perform qualified, independent audit of payments made and revenues received	Annual industry cost and fiscal audits by experienced international auditors. Auditor reconciles tax filings with revised tax assessments and with payments made, and identifies discrepancies.	Take notice and act according with the auditors' recommendations. This will allow a comparison of payments made by industry and revenues received by the federal and provincial governments. Audit revenues received by MOF, Cabinda and Zaire; include Sonangol, and clear arrears.
Publication of audit results in accessible form	Current detailed publication of company payments on MOF website. Audit results not yet published.	Add audit results. Improve accessibility of website. Consider broader media publication.
Audit exercise should apply to all companies, including Sonangol	Current practice, but Sonangol data derived from block operator.	Sonangol to provide data directly to MOF. Publication of Sonangol corporate audits
Engage civil society in revenue management and transparency process	No current engagement	Topical workshops to include civil society. Establish independent public information center
Introduce clarity on legal, contractual and fiscal framework/procedures	Legal drafts and texts difficult to access. DNI preparing tax manual	Compile and publish legal texts, procedures. tax manual.
Develop time-bound, funded, action plan for implementation of transparency agenda	No current plan, although individual components have been scheduled	Prepare and publish explicit plan

and into poverty reduction activities. However, improvements in the competitiveness of the local industry and efforts to diversify the economy remain hampered by inadequate infrastructure, poor governance indicators and less than adequate business environment. In order to address these issues there are 5 areas that will require attention:

- *Structure a dialogue between the public and the private sectors.* In the very short term, the establishment of a consultation mechanism between the government and the private sector is a critical element to secure a private-sector-led growth agenda. The private sector and the authorities should then determine together the sequencing and prioritization of the necessary reforms.
- *Reduce the number of unnecessary regulations and procedures.* Despite some recent progress in lessening bureaucratization, more needs to be done in that area. A good way to start a second phase of effective reforms could be by carrying out a systematic inventory of existing rules and regulations which affect the country's investment climate.
- *Ease restrictions on property and land registration.* It is critical that property and land registration be facilitated. The authorities could facilitate the formal registration of land by reducing both the time and the cost required to register it.
- *Revise labor regulations to introduce dynamism in the labor market.* In Angola, when an employer decides for operational reasons that s/he needs to dismiss an employee, actually doing so requires a high degree of persistence and patience. It is a costly, time consuming and cumbersome process. However, it is important that the cost of dismissal be reduced while still ensuring the protection of workers' rights.
- *Maintain government commitment to improve governance and adopt anti-corruption actions.* The World Bank estimates that a country that improves its governance from a relatively low level to an average level could almost triple the income per capita of its population in the long term, and similarly reduce infant mortality and illiteracy. In the case of corruption, research suggests that it is equivalent to a major tax that increases costs and reduces efficiency.

Options to Revamp Agriculture

Assess productions costs with a view to reduce them. In the face of an overvalued real exchange rate, the country should explore possibilities for reducing agricultural unit costs. The country has a better rainfall than many of its neighbors, has substantially lower yields, and has been cut off from technological advances in new varieties or other areas for decades. The key question, however, is whether productivity gains can be large enough to offset the disadvantages posed by the strong currency and high transportation costs. As there is no clear indication as to what is the actual cost structure of production in Angola, this question can only be answered through a case by case accurate crop budget analysis.

Increase the importance of agriculture in the budget. Increasing the priority given to agriculture and rural development through budget allocation will be an important measure to strengthen the sector. Budget allocations to the Ministry of Agriculture and Rural Development (MINADER) in 2004 reached Kz2.3 billion (approximately US$25 million). This represents only 0.64 percent of the national total budget.

Move gradually towards greater decentralization. Administrative decentralization should be encouraged as agriculture is perhaps the prototypical example of an activity that is not best managed from a capital city. The move towards expenditure decentralization, however, should be conducted gradually to avoid waste and inefficiency. Equally important concerns exist with regard to the capacity of local government structures to establish and operate the necessary administrative and budgetary mechanisms that decentralization would require.

Develop and support a market-oriented environment. There should also be a shift in policy towards further market deregulation. In the short term, this would represent adjusting and/or phasing out food aid programs to avoid unfair competition with the emerging domestic production of cereals. In the area of marketing and distribution the government is still in the business of setting wholesale and retail margins. Government controls have demonstrated in Angola and elsewhere that they result in a poorly developed trading network and poorly serviced farmers.

Improve the effectiveness of MINADER. In order to function properly and effectively contribute to the development of the agricultural sector, MINADER needs to undergo performance-enhancing changes. To increase its effectiveness, the Ministry would benefit from measures aimed at: (i) identifying redundant and unproductive positions with a view to eliminate them; (ii) reforming its structure to strengthen its extension and research services; (iii) creating a high level policy analysis unit; and (iv) strengthening the ministry's existing statistical unit.

Create better incentives to stimulate competitiveness. The development of the sector will require investments in infrastructure and a better regulatory environment. A realistic vision of the future for Angolan agriculture should include a mix of commercial and familiar farms. Both need improved transport infrastructure, a modern marketing system and a conducive regulatory environment. MINADER and provincial governments should be equipped to provide the necessary institutional support to new projects.

Improving Welfare through Service Delivery

A number of official social programs have had some success, but more needs to be done to reach the most remote and most needy areas of the country. Although existing national programs administered by the government and the private sector suggest that the intended objectives are being met, the majority of them lacks capacity to fully respond to the poor given that the demand far outstrips the supply. To increase the effectiveness of current and new social programs, the authorities should: (i) adopt mechanisms to consult the poor on how to allocate public funds; and (ii) use fiscal savings to improve service delivery. These two options are discussed below.

Reaching the Poor with Social Services

Mechanisms to consult the poor on how and where to allocate public funds can be very helpful. Angola's experience with FAS is a very good example of this mode of operation. Mechanisms to automatically consult the poor in regards to decision-making on public investment should be put in place in addition to accountability channels in order to ensure quality control and civic engagement.

Using Fiscal Savings to Improve Service Delivery

Subsidies to fuel and utility prices should be phased out gradually and the savings redirected to improve the quality of public services and to compensate the poor. In a first stage, the authorities will need to announce a comprehensive program to deal with subsidies in a phased strategy beginning in 2006. In a second stage, the current pricing mechanism used by Sonangol should be reassessed and the adjusted level of subsidies phased out gradually while the compensation program designed in the first stage continues to be implemented. At any point in time, the program should be adjusted in case there is a crash in international oil prices.

The authorities should not neglect the policy's political costs. International evidence demonstrates that fuel price hikes unaccompanied of a set of palliative measures typically leads to public protests and may trigger violence and social unrest. The experience of other countries demonstrates that any program to reallocate fiscal spending should pay close attention to the following broad principles:

- *Political attractiveness:* Off-setting, politically attractive expenditure programs are key to the political and social sustainability of a subsidy removal. A high visibility and credible announcement of compensating measures is an indispensable feature of the package.
- *Pro-poor and effective targeting:* Measures should ideally be those that maximize development and poverty reduction impact. "Compensation" programs should be targeted to benefit the poor and possibly other key losers of the subsidy removal as well. They need to be seen to do so as well.
- *Speed of spending and impact on households:* The speed at which programs are designed and money is spent is important for macroeconomic reasons, while the speed at which programs impact poor households is important for compensatory and political economy reasons.

One of the palliative measures that may bring large benefits to the poor is a water and sanitation policy. The welfare analysis carried out in the World Bank-UK-sponsored study on the impacts of phasing out subsidies indicates that changes in water price will bring about a much larger marginal social impact than changes in fuel prices. Given the precarious state of affairs in water infrastructure, management and service delivery, there appears to be ample room for reform in the public water sector with the goals of making it more efficient, expanding water supply to the population and improving water quality. The funds needed to finance such a large venture could come from funds saved with the gradual phase out of fuel subsidies and from private sector participation.

Other palliative policies that can be implemented are as follows:

- *Public Education Policy:* Promotion of voucher systems, conditional cash transfers and school meals in order to offer incentives to parents to send their children to school and incentives for children to stay in school;
- *Health Care Provision Policy:* Promote infrastructure investment, offer preventive care and mobile clinics to expand health care service access;
- *Public Transportation Policy:* Creation of a social pass which can be used in buses and vans interchangeably, cost subsidization of van service based on utilization rates, implement employer-sponsored passes.

A successful package to phase out subsidies in Angola will demand political commitment and a good implementation and monitoring strategy. In order to successfully implement the package the GoA will need to form teams and attribute responsibilities. The teams should cover the following main areas: (i) macroeconomic and fiscal issues; (ii) expenditure programming (social protection and compensatory expenditure, and other development and poverty reducing expenditures); and (iii) socialization of savings.

Finally, in order to know if a poverty reduction strategy is effective in reducing poverty, it is necessary to set in place a poverty monitoring system to track key indicators over time and space. In Angola, there is still a dearth of knowledge on the livelihoods of the poor. Overall, it is difficult to have a full assessment of the existing safety net programs given that these do no have at the outset an expected number of beneficiaries. This failure comes from the lack of reliable data on poverty to which programs could refer to in order to assess appropriateness and effectiveness in their targeting and the impact of their activities. A sensible way of tackling this limitation would be to strengthen the National Statistical Institute (INE) and establish a program of annual monitoring of poverty and indicators and disseminate these to the public at large.

Summary of Recommendations and Next Steps

The following Table outlines priority areas for action and suggested next steps (see Table E.2). Some recommendations are already under implementation, in some cases with World Bank support.

Table E.2. Summary of Diagnostics and Recommendations

Reform Areas	Recommended Next Steps
1. Strengthen Macroeconomic Performance: *The main objective is to strengthen economic policies to guarantee sustainable macroeconomic stabilization and insulate the economy from volatility associated with international oil prices (Chapters 1, 2, and 3).*	
Address continuing deficiencies in macroeconomic policy design and implementation.	Adopt a prudent fiscal envelope for the coming years and avoid rapid increases in public expenditure. Continue adopting conservative oil price in the budget.
	Prepare an MTFF. Calculate annual ceilings for public expenditure based on medium-term revenue forecasts evaluated at a long-run oil price. Move gradually towards an MTEF.
	Create clear fiscal rules for the recently announced oil reserve account based on solid governance and transparency principles. Adopt a cap ex-ante limiting the volume of resources to be accumulated.
	Avoid pro-cyclicality of fiscal spending. Generate fiscal surpluses when oil prices are rising to sustain spending when revenues eventually fall.
	Use oil revenues for productive investments and not consumption (increase productive capacity) to reduce costs to the entire economy.
	Adopt a clear monetary anchor, use foreign exchange sales to mop liquidity and keep the downward pressure on inflation, and do not resist likely appreciation of the currency.
	Build up foreign exchange reserves and reduce the external debt.
	Modernize policies and institutions for procurement, financial management, and program evaluation.
2. Address Outstanding Issues on Governance and Transparency: *A sound strategy to manage the country's mineral wealth requires sound governance and transparency in the oil and diamond sectors to avoid rent-seeking and waste (Chapters 3 and 4).*	
Deal with conflicts of interest in Sonangol and Endiama.	Pursue adequate government oversight of Sonangol and Endiama.
	Separate the activities of Sonangol and Endiama as operators and regulators in their respective sectors by transferring regulatory responsibilities to appropriate ministries.
Improve governance and transparency in the oil and diamond sectors.	Support the petroleum revenue management unit in the Ministry of Finance and train staff to work with the oil revenue forecasting model.
	Publish Sonangol's and Endiama's corporate audits.
	Define a plan and a timeframe to phase out completely Sonangol's quasi-fiscal operations.
	Update information on oil exports, oil prices, and profit oil without delays on the website of MINFIN.
	Endorse formally the EITI criteria.
	Compile and publish legal texts, procedures, and tax manual on fiscal/legal framework in the oil sector. Enact transparent and stable legislation under a new diamond law.

(Continued)

Table E.2. Summary of Diagnostics and Recommendations (*Continued*)

Reform Areas	Recommended Next Steps
Improve the business environment in the diamond sector.	Commission an updating of the 2003 Diamond Diagnostic Study.
	Review and regularize existing mineral rights (exploration and mining).
	Produce better geological information and disseminate it.
	Liberalize diamond sales; identify and assess options for free market trading of diamonds; pilot test selected options for free market trading of diamonds.
	Raise awareness and disseminate rules of the game locally and internationally.
	Reinforce implementation of the Certification of Origin Scheme.

3. Promote Private Sector Development: The climate for doing business in Angola is perceived as one of the least conducive in the world and any strategy to support private sector development will need to promote pro-business reforms (Chapter 5).

Remove barriers and constraints to the business environment.	Structure a dialogue between the public and the private sectors
	Reduce the number of unnecessary regulations and procedures.
	Ease restrictions on property and land registrations and reduce the associated costs.
	Revise labor regulations to introduce dynamism in the labor market.
	Maintain government commitment to improve governance and adopt anti-corruption actions.
Use trade as a yardstick for measuring success in diversification.	Resist protectionist pressures that will emerge as the exchange rate appreciates.
	Develop export incentive systems where there are comparative advantages.
	Improve trade facilitation and strengthen trade capacity.
	Take advantage of trade agreements and participate actively in regional and global negotiations.
Introduce win-win schemes to foster interaction between oil companies and local suppliers.	Prepare sector development strategies.
	Invest in training for Angolan workers.
	Promote joint ventures.
	Improve business environment, competition, and exit mechanisms.

4. Strengthen Agriculture Sector: Agriculture can play a significant role in the future of Angola but this potential can only be realized through much needed institutional reforms and a higher priority to the sector in the state budget (Chapter 6).

Increase competitiveness in agriculture.	Review agricultural support policies (trade and fiscal) with a view of formulating a clear strategy to increase the competitiveness of the sector.
	Support smallholders and resist the temptation to engage in public production or market activities.
	Repair roads in rural areas to facilitate access to markets.

(*Continued*)

Table E.2. Summary of Diagnostics and Recommendations (*Continued*)

Reform Areas	Recommended Next Steps
Remove obstacles to output growth	Prepare a detailed analysis of actual cost structures per crop with a view to reduce production costs.
	Increase the importance of agriculture in the budget.
	Move gradually towards administrative and spending decentralization.
	Develop and support a market-oriented environment for the sector; review existing marketing controls; phase out existing government involvement in production and marketing activities; review the current policy of distribution and subsidized sales of fertilizers.
Improve the effectiveness of MINADER	Perform an overall manpower study of MINADER with a view to identify redundant and unproductive positions and formulate a plan for their elimination.
	Reform the structure of MINADER to strengthen its extension and research services.
	Create a high level policy analysis unit in the Ministry capable of analyzing policy alternatives and appraising projects.
	Improve the existing statistical unit in the Ministry that should liaise with the National Institute of Statistics and the Ministry of Planning to design a national agricultural survey.

5. Improve Public Service Delivery: Improve living standards through government spending and the delivery of social and economic services (Chapter 7).

Design a phased public works program.	Target public works jobs program for the low skilled labor force.
Increase job opportunities for the labor force.	Revise labor legislation to introduce dynamism in the labor market (but respect workers' rights).
	Support the organization of small business into cooperatives.
Encourage greater community participation in the design and implementation of social service investments.	Adopt best practice mechanisms to consult the poor in regards to decision-making on public investment.
	Increase scope and coverage of the FAS.
	Promote a higher degree of coordination among development partners and the government (local and provincial).
Use savings with subsidies phase out to improve service delivery.	Prepare a phased program to eliminate fuel and utility price subsidies. Adopt a two-pronged strategy to: (i) bring fuel prices subsidies bill in line with the oil price implicit in the 2006 budget; and (ii) revise Sonangol's pricing mechanism to phase out completely fuel subsidization and liberalize distribution.
	Use fiscal savings to compensate the poor and the losers with the elimination of the subsidies by investing in infrastructure rehabilitation in the water sector; public education; health care provision; and public transportation.
	Create an inter-agency team to prepare the package. The team should look at the following areas in designing the package: macroeconomic and fiscal issues; expenditure programming (social protection and compensatory spending); and socialization of savings.

(*Continued*)

Table E.2. Summary of Diagnostics and Recommendations (*Continued*)

Reform Areas	Recommended Next Steps
Establish a program of annual monitoring of poverty and social indicators	Strengthen the National Statistics Institute (INE) with better physical and human capital. Higher wages may be necessary to attract skilled labor.
	Revise the national statistics plan and give a higher priority to poverty monitoring.
	After adequate support is in place, conduct and disseminate annual labor force surveys.
	Frame public spending plans within a medium-term strategy for poverty reduction and for achieving the Millennium Development Goals (MDGs).

Introduction

The one and only Country Economic Memorandum about Angola was prepared and discussed with the authorities in June 1990 (World Bank 1990). At that time, the Bank's diagnostic about the Angolan economy identified three main factors that contributed to slow economic growth and to continuous economic instability since Independence in 1975:

- First, the violent conflict that erupted with Independence and that contributed to the destruction of infrastructure and to the exodus of a huge number of people into the cities. The conflict disrupted internal transportation, imposed a heavy burden on the budget (with defense accounting for more than 40 percent of government expenditures at the time), absorbed a large proportion of the limited supply of technicians and skilled manpower, and created enormous suffering and deprivation among the population.
- Second, the massive exodus during the period of transition to Independence of about 300,000 Portuguese settlers (90 percent of the total) who held practically all administrative, managerial and skilled jobs, which created a situation of chaos in the economy. Few Angolans had the professional qualifications to run the enterprises which were abandoned or to fill the jobs which had been left.
- Third, the inefficient economic management and inadequate economic policies which prevailed after Independence which, along with the previous two factors, contributed to the decline in aggregate production, scarcities in the supply of consumer goods and inputs for the industry, and distortions in the distribution of income.

The World Bank's advice to the authorities to put the economy back on a path of macroeconomic stability and equitable development reinforced the current economic policy thinking at the time, which was based on the Government's own reform agenda in the form of a Program of Economic and Financial Restructuring (designated as SEF, *Programa de Saneamento Económico e Financeiro*). The main recommendations were oriented towards the following two main broad objectives:

1. Stabilization of the financial situation, by reducing internal and external disequilibria, which were reflected in inflationary pressures, large budgetary deficits, excessive losses and indebtedness of many enterprises, serious deterioration of the financial situation of the banking system, accumulation of arrears in foreign payments, and difficulties in servicing the external debt;
2. Introduction and implementation of structural reforms, in order to increase productivity, improve the allocation of resources and create the conditions for a faster rate of economic growth and equitable development in the future.

Some 15 years later, the main issues confronting the Angolan authorities in their efforts to consolidate macroeconomic stability on a sustainable basis and in promoting an improvement in the welfare of the Angolan citizens do not seem to differ significantly from those addressed in the 1990 report. Therefore, in the current Country Economic Memorandum, the Bank reassesses some of the key issues that remain relevant nowadays and that should help the Angolan economy reach a path of sustainable economic development.

The analysis in this report centers around the following four core issues: (i) taking stock of socio-economic realities; (ii) the options available for the management of the country's mineral wealth without deleterious macroeconomic consequences; (iii) the main constraints to economic diversification away from the mineral sectors; and (iv) the challenges and opportunities to improve the welfare of the population. Each of these core issues forms the building blocks that provide an overview of the current situation and a possible solution to Angola's structural problems in the short to the medium term. The report thus plays an informative role and offers policy recommendations. The rationale behind the core issues and how they fit in the structure of the report is briefly outlined below.

Taking Stock of Socio-Economic Realities. The report starts by taking stock of socio-economic realities in the country that may help to understand how its institutions developed and how the Government has handled macroeconomic management since Independence. It traces the link between the recent history of conflict, weak institutional capacities, and the difficulties in achieving macroeconomic stability after Independence. This is done in Chapters 1 and 2. In Chapter 1, the analysis starts with a brief discussion of socio-economic realities in the country. In Chapter 2, a comprehensive macroeconomic assessment is presented highlighting major past features, the country's constant search for stability, and recent successes in the macroeconomic front.

Options to Manage the Mineral Wealth. With rising international oil prices and the prospects of a doubling in oil production in the near future, Angola will benefit from a sizable revenue windfall. The prospects in the diamond sector are equally promising. The significant additional revenue and the speed with which the government will have access to it poses challenges of macroeconomic management and raises intergenerational considerations and expectations about how the mineral wealth can contribute to improve the welfare of the poor and of the vulnerable in Angola. These questions are addressed in Chapters 3 and 4. In Chapter 3, the report discusses the structure of the petroleum sector, the future production profile, the size of the oil wealth, and policy options to manage the revenue windfall. Chapter 4 focuses on the diamond sector, its structure, legal and fiscal framework, and explores ways in which the sector can improve its contribution to social development.

Constraints to Economic Diversification. Despite the favorable outlook in terms of mineral wealth, the Angolan economy will not enter a path of sustainable shared growth without some necessary structural reforms. Because of the long civil war and of the effects of the strong dependence on oil and diamond revenues, the private sector has not evolved outside of the mineral sectors and the quality of the country's institutions remains low. The combination of this state of affairs creates constraints to private sector development and for the diversification of the economy. Options to overcome these structural issues are discussed in Chapters 5 and 6. In Chapter 5, the report assesses the quality of the business environment and the opportunities to improve the investment climate. Chapter 6 discusses alternatives to unleash the potential of the agricultural sector in generating employment outside of the mineral sectors.

Challenges and Opportunities to Improve the Welfare of the Population. Finally, the report acknowledges that it will take some time before the required structural reforms can start to yield concrete results. In this instance, the authorities will need to worry about mitigating measures to improve the welfare of the population until the Angolan economy reaches a path of sustainable development. In Chapter 7, the analysis focuses on how to improve the livelihoods of the poor and of the vulnerable with recommendations on how to use the mineral wealth to improve public service delivery targeted to the poor. The analysis in Chapter 7 is complemented by an additional study on the welfare impacts of phasing out fuel and utility price subsidies and on how to use the derived fiscal savings to improve service delivery in Angola. This study, that was managed by the World Bank, funded by the British Government, and prepared in the context of this Country Economic Memorandum, is published separately.

As policy reform in an economy in transition (from war to peace, from a colonial economy to an independent market system, and from Marxist centralization to a multi-party democracy) is by definition politically difficult and requires a piecemeal approach, the report presents policy options that range from the modest to the radical. The recommendations include institutional reform but, recognizing that this will be difficult and time-consuming, other alternatives are suggested. To guide policymakers, the report also attempts to determine priorities and sequencing. This is an issue that surfaces in all chapters when policy recommendations are offered.

A note of caution is necessary in relation to the quality and completeness of the data used in the preparation of this report. The reader is alerted that in many occasions the necessary information was either not available, or when available was of limited coverage, quality and usefulness. A great deal of this is associated with the impacts of the conflict in the capacity to generate data. The Government now has a long-term plan to address this problem and it is hoped that data quality and coverage in Angola will improve substantially in the near future.

CHAPTER 1

Country Background: Socio-Economic Realities Before and After Independence

Angola is a country blessed with vast stocks of natural resources, of which oil and diamonds alone account for more than half of the country's gross domestic product and for over 90 percent of its exports. The economic concentration on oil and diamonds, together with the deleterious effects of a violent war that plagued the country for nearly three decades, has left a difficult legacy. The potential to develop a thriving diversified economy outside the mineral sector is large, but so are the obstacles that have to be faced before the country can take advantage of the favorable moment. This Chapter takes stock of the socio-geographic characteristics of the country and of the structural changes that have taken place since Independence and sets the stage for the analysis that follows which will deal with the challenges of achieving sustainable macroeconomic stability while promoting shared-growth.

Socio-Geographic Characteristics

The Republic of Angola is, after the Democratic Republic of Congo and Sudan, the third largest nation south of the Sahara. It has an area of 1,276,700 sq. km (including the 7,270 sq. km of the oil-rich Cabinda enclave) and is the largest Portuguese speaking African country. Angola is located on the West Coast of Africa bordering Namibia, the Democratic Republic of the Congo, and Zambia. Without considering Cabinda province (an exclave in the Northwest separated from the rest of the national territory), Angola has a roughly square shape, measuring 1,277 km from North to South and 1,236 km from West to East (from the mouth of the Cunene river to the Zambia border). Angola's capital, Luanda, lies on the Atlantic coast in the northwest of the country. Plateaus, averaging altitudes between 1,050 and 1,350 meters, account for about two thirds of the Angolan territory. The Angolan coast, with an extension of 1,609 km, is mountainous to the North of the mouth of the Kwanza river, and quite flat with occasional cliffs to the South. The coastal plains are separated from the inland plateau by a score of irregular "terraces" that form a subplateau. The most important rivers in the country originate in the plateau regions and flow in three directions: east-west to the Atlantic, south-southeast, and northeast. Most rivers, however, do not provide easy access to the interior regions as they are not navigable. Nevertheless, they offer energy and irrigation potential. The main rivers of Angola include the Kwanza (with an extension of 960 km, 200 of which are navigable by small watercraft), and the Cunene (945 km long and bordering Namibia to the south).

The country's relative climatic diversity represents an advantage and hints at huge potential for agricultural development. Angola's location in the intertropical and subtropical zones of the Southern hemisphere, its proximity to the sea and the cold Benguela stream, and its topographical characteristics are the factors which create two distinct climate regions with two seasons: the dry and cool season (from June to September) and the hot and humid season (from October to May). The northern region from Cabinda to Ambriz has a humid tropical climate with heavy rainfall, while the region from Luanda to Namibe (Moçamedes) has a moderate tropical climate, with the rainfall reduced on the coast by the Benguela wind stream. The southern strip between the plateau and Namibia has a desert climate, given the proximity to the Kalahari, with irregular rainfall between 600 and 1000 mm. annually. Temperatures average 23°C in the north and the coastal areas, and 19°C in the interior. The relative climatic diversity, due to variations of altitudes across the country, allows for the growth of crops from both tropical and relatively more temperate zones.

The vast and diverse territory hosts a large concealed economic potential. Among the abundant natural resources there is plenty of water that provides for hydroelectric power plants and irrigation; amid mineral resources there are abundant oil, diamonds, iron, quartz, ornamental stones and phosphates. In the Cabinda region, very dense forests predominate (Maiombe forests) with economically important timbers such as black wood, ebony, African sandalwood, and ironwood. With a coastline of 1,650 km, Angola's waters are rich in fish, mollusks, and crustaceans. The main petroleum basins under exploration are located near the coast of Cabinda and Zaire provinces. The main diamond producing area is located in the Lunda Norte province. Unfortunately, due to the nonexistence of proper and comprehensive geological surveys, the whole mineral potential of Angola is, to this date, vastly unknown.[5]

The Angolan population is young and is growing rapidly. Recent population figures are difficult to obtain due to the lack of a full national census. A limited census was carried out in the province of Luanda in 1983, which was extended to the provinces of Cabinda, Namibe and Zaire in 1984. War-related problems made it impossible to carry out a national census. It took 70 years for the population to double from 2.7 million in 1900 to 5.9 million in 1970, with the rate of growth accelerating in the 1940–70 period, due to significant Portuguese immigration. In 1980, according to official estimates, the population reached 7.7 million, implying an annual average growth rate of 3.2 percent for the previous decade. Though data is scanty, the population was projected to grow at an annual rate of 2.9 percent during the 1980s and 1990s, reaching about 13 million in 2003. Estimates by

---

5. See Araújo and Costa (1997) and Alves da Rocha (2001) for a detailed description of Angola's natural resource endowments.

the U.S. Census Bureau suggest that in 2000 some 6.5 million people or about 62 percent of the Angolan population were under the age of 24 and that by 2025 that segment of the population would be of approximately 10.8 million people, or around 60 percent of the population. With a population growth rate of approximately 2.9 percent, the absolute numbers and the proportion of youth will continue to be extremely high over the next 50 years.

Population density is low and ethno-linguistic groups are geographically separated. Population density (8.6 inhabitants per sq. km) is very low, with the most populous provinces being Huambo, Luanda, Bie, Malange, and Huila, which together account for more than half of the total population. About three-quarters of the population come from three ethno-linguistic groups: the Ovimbundu (37 percent) in the Central plateau region, the Kimbundu (25 percent) living in a belt extending from Luanda to the East, and the Bakongo (13 percent) in the Northwest. In addition, *mestiços* (Angolans of mixed European and African family origins) amount to about 2 percent, with a small population of whites, mainly ethnically Portuguese. Portuguese is both the official language and predominant language, spoken in the homes of about two-thirds of the population, and as a secondary language by many more.

The historical ethnic divide influenced the anti-colonial resistance and the formation of the post-colonial state. The Ovimbundu are, by far, the largest ethno-linguistic group. They dominate the areas with the highest population density in the country—the central plateau provinces of Benguela, Bie, and Huambo. As argued by Malaquias (2000), their cultural, linguistic and economic domination in central part of Angola is such that they have been regarded in the past as "a nation rather than an assembly of tribes." Awareness of this ethnic diversity is crucial to understanding the politics and society in Angola, both during colonial times and as a post-colonial state. For example, in the past, the Portuguese were able to impose colonial rule because the nature of anti-colonial resistance was so fractured. Although sporadic military resistance persisted during Portugal's presence in Angola, the various kingdoms and chiefdoms threatened by colonial domination were not able to create a united front. From this perspective, the disunity that characterized the anti-colonial movement and the inability to establish an inclusive political system after independence have long historical antecedents.

Living Standards Indicators

There is a dearth of information on poverty and social indicators and most of the existing surveys are dated and curtailed in terms of coverage. The most recent household income and expenditure survey (IDR) is from 2001 and covers only eight provinces, about 50 percent of the population. The survey covers rural and urban households, but owing to the provinces covered, 86 percent of the sample lived in urban areas. Although Angola was highly urbanized at this time, as many households fled the countryside and moved into displaced person camps in urban areas, this sample is considered to have an urban bias. Female headed households are also likely to have been undercounted, as only 20 percent of households were female headed in the sample, but owing to the effects of the war, the proportion is believed to be higher. In 2002, a Multiple Indicators Cluster Survey (MICS) was conducted and collected data on household assets, health and education status, and access to and use of social services. This survey covered 19 provinces, approximately 90 percent

Table 1.1. Basic Poverty and Social Indicators

Indicator	Where Angola Stands
Population (million)	14.7
Population ≤20 years	60%
Population below poverty line	68%
Life expectancy at birth	42.4
Under-five mortality (per 1000 live births)	250
HIV/AIDS prevalence	3.9%
Population who know where to get an HIV test	23%
Population correctly stating 3 main ways to avoid HIV infection	17%
Adult illiteracy rate	33%
Maternal mortality rate	1800
Net primary school attendance rate (1-4th grade)	56%
HDI rank (out of 177 countries)	166
GDP/capita rank (out of 177 countries)	128
Gini coefficient (income, 1995)	0.54
Gini coefficient (income, 2001)	0.62

Source: IDR (2000/1); UNICEF (2003); UNAIDS (2004); UNDP (2005:1, 5).

of the country. This sample is considered more reliable. The MICS results show that 61 percent of sample households reported living in urban areas, but this probably still undercounts the rural areas.

The existing social indicators still reflect conditions prevailing during the war period. According to both the 2001 IDR and the 2002 MICS, approximately 70 percent of the population live on less than 2 dollars a day and the majority of the Angolans lacks access to basic healthcare. About one in four Angolan children die before their fifth birthday, 90 percent of whom perish due to malaria, diarrhea or respiratory tract infections, the maternal mortality rate (at 1,800 per 100,000 births) is one of the highest in SSA, and three in five people do not have access to safe water or sanitation. The HIV/AIDS prevalence rate is, according to official statistics, relatively low, affecting an estimated 3.9 percent of adults.[6] However, lack of statistical information and a limited number of surveillance centers suggest that the true prevalence rate may be much higher. In terms of education, primary school enrollment is very low at 56 percent, and suffers from late entries into school and high repetition and drop out rates. Some 33 percent of the population is currently illiterate, though in rural areas this climbs to as many as 50 percent (see Table 1.1).

Living standards have shown some improvement recently, but they continue to be considered low by international standards. The crude birth rate of 52 per thousand estimated for 2000-05 is the second highest in the world after Nigeria and reflects a high total fertility rate (6.8) combined with a large proportion of women of fertile age. The crude

---

6. UNAIDS (2004). Low estimates are 1.9% and high estimates are 9.4%.

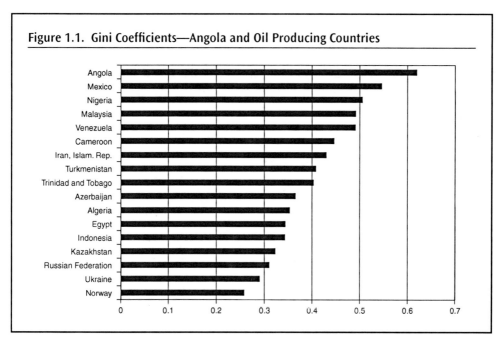

Source: 2005 World Development Indicators.

mortality rate of 25.9 per thousand for 2000-05 is influenced by malnutrition, precarious sanitary conditions, a large proportion of illiterate mothers, and inadequate health facilities. Infant mortality is very high, although precise information is still difficult to obtain, ranging from 191 to 250 per thousand for the period 2000–05. Life expectancy has increased from 35 years in 1960–70 to 42.4 years in 2000–05, but remains very low. The rapid growth of the population, together with a relatively low life expectancy, is a clear indicator of a young population. The proportion of the population under 20 years of age is 60 percent, whereas those 65 years or older represent only 2.8 percent. As a result, the dependency rate is extremely high. Additionally, according to the UNDP's 2005 Human Development Report, Angola is among the group of least developed countries in the world with a human development index (HDI) of 0.445, which ranks it at 166 out of 177 countries. The vast majority of the population (68 percent in 2001, according to official data) lives below the poverty line. Furthermore, the household budget survey (IDR) of 2001 shows that income inequality in fact rose throughout the 1990s, from a Gini coefficient of 0.54 to 0.62, making Angola one of the most unequal countries in the world, in terms of income distribution (see Figure 1.1).[7,8]

---

7. A recent study by Adauta de Sousa and others (2003: 38) estimates that the national Gini coefficient is actually higher than the 0.62 recorded for Luanda.

8. Evidence presented in the 2006 World Development Report suggests that higher initial inequality means that growth reduces poverty by a lesser amount as it would otherwise do in the absence of income inequality (p. 86).

Inequality is high in terms of asset holdings which are mostly found in urban rather than in rural areas. Using the asset scale estimated with using the MICS data (the MICS did not collect information on consumption or income), assets are found to be highly concentrated in urban areas—only 8 percent of the population in urban areas was found in the lowest quintile according to asset and durable goods holdings, but 44 percent of the rural population was in this quintile.[9] Likewise, only 2 percent of the rural population was in the upper quintile. Households with larger asset holdings are much more likely to be found in the capital region, followed by the Western and Southern regions, although the latter region also has one of the highest concentrations of low asset holders. This region is very heterogeneous, as Cunene province has 41 percent of its population in the lowest group, while Namibe has the same amount in the highest quintile. The assets most likely to be held are radios (76 percent in Luanda, 28 percent in rural areas) and bicycles (9–10 percent outside of Luanda). Few have cars outside of Luanda.

Inequality of opportunities is also stark. Data from the 2002 MICS survey suggest that the ability to gain access to basic social services is directly influenced by the household's income level. Access to primary education, for example, is only 35 percent for the poorest quintile, while for the richest quintile it is more than double that, at 77 percent.[10] Literacy rates compare at 62 percent for the poorest quintile and 95 percent for the richest quintile. Gender disparity is also apparent, with female literacy rates within the poorest quintile of 27 percent compared with 62 percent for men, and 86 percent and 95 percent within the richest quintile, respectively.[11] Government data from 2002 show that overall completion rates for primary education for girls are estimated at 41.3 percent compared with 56.8 percent of boys.[12]

General living conditions are far from ideal, even for the middle class, but they are especially dire for the poor. The long period of civil war destroyed much of the infrastructure. Most Angolans, even in urban areas, do not have reliable access to safe water—only 20 percent in urban areas outside of Luanda have access to it, according to the MICS data. Again the inequality is stark.[13] In Luanda, virtually no one in the two lower assets quintiles

---

9. Unequal access to assets and services is an important to constraint market efficiency, as failures in the market for credit, insurance, land and human capital result in underinvestment by the poor, overinvestment by the rich and a less efficient economy (see WDR, 2006, Chapter 5).

10. Conditional cash transfers (CCTs) have been used in many countries to induce poor parents to enroll their children in school. CCTs make payments to poor families, typically mothers, on the condition that children attend school regularly. The programs can be seen as compensating for the opportunity cost of schooling for poor families and represent one approach to addressing failures in credit markets and the imperfect agency of parents. The biggest programs of this kind are Oportunidades (formerly PROGRESA) in Mexico, the Bolsa Escola in Brazil, and the Food for Education Program in Bangladesh (see WDR, 2006, p. 137).

11. The 2006 World Development Report argues that gender inequity directly affects the well-being of women and decisions in the home, affecting investments in children and household welfare (p. 51). Econometric evidence cited in the report confirms that an increase in a woman's relative worth and an improvement in her fallback options have positive effects on the children (p. 53).

12. Progress report on MDGs (2005: 24). Gender inequities are also reflected at the government and national political level, where women only account for 11 of 70 ministers and vice-ministers and only 14% of the national assembly. At the Provincial level, there are currently no female Governors or Vice-Governors.

13. According to a recent survey conducted by Development Workshop in Luanda, about 30% of the interviewed households did not have access to basic infrastructure (e.g. piped/safe water and electricity), as well as to basic services such as health and education in the vicinity. About 56% have access to some level of infrastructure and services. Only 13% have access to a relatively high provision of infrastructure and services (Development Workshop, 2003: 44).

reported access to safe water, while 40 percent in the highest quintile reported access. For example, in the *comuna* of Hoji ya Henda, of 580,000 people, only about 15 percent of the people are connected to piped water while the rest of the population relies on 18 public water points, 14 of which are functioning. Electricity also is primarily available to the rich, most of whom rely on generators given the poorly functioning infrastructure and frequent power outages. In Luanda, 82 percent of the highest quintile reported having electricity, but no one in the bottom 60 percent reported having any.

Public delivery of social services is also skewed in favor of the urban rich. For example, in urban areas in 2002, 50 percent of women reported receiving some form of trained pregnancy assistance, and this percentage only dropped to 40 percent in the bottom quintile. However, only 24 percent in rural areas reported receiving this assistance, with only 16 percent in the poorest 20 percent of the population. An estimated 17 percent of (surviving) children under the age of 5 had not received any childhood vaccinations at all in 2002. Access to education is poor as well. Only 44 percent of rural children of primary school age (grades1-4) are reported to be in school, and 60 percent of urban children. This is partly because about one third of children start school 1–2 years late, either because the walk to schools is long, or the family cannot afford the fees[14], or they are needed to work at home, or the quality is viewed as poor so parents do not value the child's education very highly, or the parents simply want to keep the children at home for an extra year. Of those who start, only 46 percent complete primary school and enroll in fifth grade.

The Transition to Independence

Rapid economic growth developed after WWII and was largely based on agricultural exports. In the early decades of the 20th century, there had been some development of the modern sector of the Angolan economy based on railway building, diamond mining, plantation agriculture and trade. However, rapid economic growth began only after the Second World War. The initial stimulus came from the coffee boom. Coffee production rose from 14,000 tons in 1940 to around 100,000 tons in the early 1960s. In 1950, coffee was already the most important of Angolan exports, accounting for 30 percent of total foreign currency earnings. Widening economic opportunities contributed to an increase in the population of Portuguese settlers, from 44,000 in 1940 to 172,000 in 1960. However, most Angolans continued to live in extremely difficult conditions, many of them became employed in the plantations and mines, either on a voluntary basis (at very low wage levels), or under the so-called system of contract labor, which imposed forced labor on about 350,000 workers by the mid-1950s and that was only abolished in 1961.

Heavy public investment in economic infrastructure boosted economic growth between 1960 and 1974. Economic growth accelerated considerably in the period 1960–74, despite the anti-colonial war of independence that started in 1961. Public investment in economic infrastructure increased significantly, restrictive investment laws were liberalized in 1965 to encourage foreign investment and the Portuguese authorities also encouraged

---

14. In the 2001 IDR, 25% of parents reported that their children were not in school owing to lack of money to pay fees.

the immigration of Portuguese settlers. During the period 1960–74, GDP rose at an annual growth rate of almost 7 percent in real terms, one of the highest in Africa.

The volume of annual coffee production doubled to 210,000 tons and by the early 1970s Angola ranked fourth among the world's coffee producers. Besides coffee, several other cash crops (such as sisal, sugar, tobacco, and cotton) contributed to foreign exchange earnings. In the early 1970s, Angola was also the fourth largest producer of diamonds (over 2 million carats a year), and the production of iron ore, which was negligible in the 1960s, exceeded 6 million tons per year in 1970–73. The most spectacular development, however, was that of oil production. The first commercial discovery of oil resources was made in 1955 and production rose rapidly in subsequent years, particularly after 1969. In 1973, output was already in the order of 144,000 barrels per day and, following the price increases in that year, oil overtook coffee as the leading export commodity, accounting for more 30 percent of total export revenues.

Economic development under colonial rule did not necessarily benefit the bulk of the Angolan population. In some areas, the best land was taken away from the Angolans and given to Portuguese settlers, and the compulsory relocation of large numbers of African peasants into protected villages (as a counterinsurgency measure) was also very disruptive of traditional African agriculture. Educational standards were very low and Africans were denied access to education. As a result, native Angolans were not only absent in managerial, professional, and technical employment, but due to the influx of poorly educated Portuguese settlers, almost all semi-skilled jobs and a high proportion of low skilled jobs were reserved for European immigrants.

The armed struggle for independence of Angola from Portuguese colonial rule began in 1961. The hostilities started in 1961 and were conducted by three rival movements: the MPLA (*Movimento Popular para a Libertação de Angola*), the FNLA (*Frente Nacional de Libertação de Angola*) and UNITA (*União Nacional para a Independência Total de Angola*). A separatist movement also appeared in the Cabinda region (*Frente para a Libertação do Enclave de Cabinda*—FLEC). A number of commentators sustain that UNITA emerged because of a perceived dominance of the MPLA and the FNLA by mixed-race intellectuals from the coastal cities.[15] UNITA leaders claimed to represent the "real Africans", sons of the soil, living in the bush, fighting against a wealthy, cosmopolitan, better-educated urban elite.

Ethnic tensions permeated the conflict for liberation from colonial rule and fueled a civil war after independence. Although other important dividing factors, including class and race, were superimposed by colonialism on Angolan society, the perceived gap dividing different ethnic groups proved to be both enduring and difficult to bridge. Not surprisingly, the anti-colonial struggle reflected these ethnic differences, as the three major liberation movements—MPLA, UNITA, and FNLA—represented almost exclusively the Mbundu, Ovimbundu, and Bacongo ethnic groups, respectively. As argued by Malaquias (2000), even in the face of their common enemy, Portuguese colonialism, these three nationalist groups were unable to overcome their differences in the form of a united front. The independence of Angola was officially proclaimed on November 11, 1975, with the MPLA in control of the government. Despite the proclamation of independence, UNITA

---

15. See, for example, the discussion of the roots of the Angolan war in the book by Dietrich and Cilliers (2000).

maintained a guerrilla war in the South, with the military support of South Africa and others. This war was enlarged in subsequent years creating extremely severe disruptions to the economy.

The war caused the demise of the rural economy and the subsequent sharp rise in urbanization due to the arrival of rural refugees. More than 1 million lost their lives during the civil war, 3 million fled to the cities and 400,000 crossed the borders into neighboring countries. Upwards of 45 percent of the population became concentrated in urban areas, with more than half of them in Luanda (Adauta de Sousa, 2003). Furthermore, the current population growth at 2.9 percent per annum has almost doubled the population since 1980, which is now estimated at 14 million. Cross-continental transportation links, which served landlocked neighbors as well as the domestic economy, have atrophied. Infrastructure has also deteriorated in the cities, partly through warfare and partly because inefficiencies in most parastatal companies and price control policies depress public utility revenues, which fail to recover costs in most services. An estimated $4 billion may be required just to restore the road and bridge network, without which little rural activity is feasible.

The structure of the economy changed significantly and the country is now highly dependent on oil. As a colony of Portugal, for nearly 500 years Angola served the needs of Portugal. The cycles of the colonial economy were determined by exports—first slaves, then primary commodities such as rubber and coffee. Before Until 1975, the country was known as an agricultural producer, not an oil exporter. It was the world's fourth-largest exporter of coffee and one of the largest exporters of staple foods in sub-Saharan Africa—exporting more than 400,000 metric tons of maize annually. These grain exports were produced almost exclusively by smallholders using traditional technologies. Oil had not yet achieved the high production levels of the 1980s and thereafter. Today, the economy is heavily dependent on oil, a capital-intensive sector with few linkages to other parts of the economy and little impact on employment. After 1973, the structure of the economy changed substantially as the mining and service sectors increased their share in GDP (Table 1.2).

Table 1.2. Composition of GDP by Sector, 1966–2004

	1966	1970	1987	1996	2004
Agriculture, Forestry and Fishery	14.2	9.0	12.6	7.0	9.1
Industry	22.2	29.6	57.5	67.8	58.1
Mining	6.3	10.7	51.0	61.2	49.8
Manufacturing	8.7	10.7	3.7	3.4	4.2
Electruicity and water	0.9	0.9	0.3	0.0	0.0
Construction	6.3	7.3	2.5	3.1	4.0
Services	63.6	61.4	29.9	25.2	32.8
Transport and communications	6.3	5.9	2.7	0	0
Commerce	34.0	30.3	7.2	15.0	15.4
Other services	23.3	25.2	20.0	10.1	17.5

Sources: IV Plano de Fomento 1974–1979, Angola; Perfil Estatistico, 1988–1991; "Angola: An Introductory Review," The World Bank, January 1991; data proviced by Angolan authorities to IMF and WB.

To this date, the Angola economy remains heavily dependent on the oil sector, a capital-intensive sector with very few linkages to other sectors of the economy and little impact on employment.

Policy Choices and Structural Changes

Important structural changes have taken place in Angola since Independence. Over the last 40 years or so, the Angolan economy suffered with adverse shocks associated with changes in international oil prices and with the consequences of the war that, together with the political orientation chosen by the Government since 1975, help to explain the structural changes which the economy has experienced since Indpendence. Following the end of the colonial period, for example, there was an almost immediate shift in public-private roles in the economy which implied in changes in the political governance and administration systems. This shift preceded changes in the composition of production and in the structure of trade. In addition, with the focus on the role of the state as the engine of growth, there was a structural change in the composition of public outlays after 1975. The next paragraphs explore these structural changes in more detail.

The economy experienced a great deal of ups and downs in its growth path during the last four decades. From 1960 to 1973, GDP per capita at 1996 international prices grew steadily, but collapsed by more than 35 percent after independence (see Figure 1.2). The period between 1974 and 1976 and the events associated with the fight for independence had a profound impact on Angola's economy insofar as skilled labor fled the country and organizational capacity deranged. From 1975 to 1997, the economy suffered several shocks, the biggest of them being the restart of the war at the end of 1992 which caused another

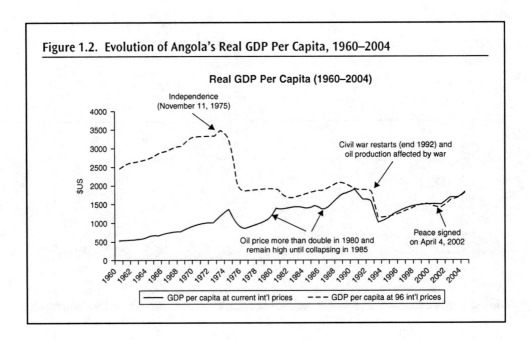

Figure 1.2. Evolution of Angola's Real GDP Per Capita, 1960–2004

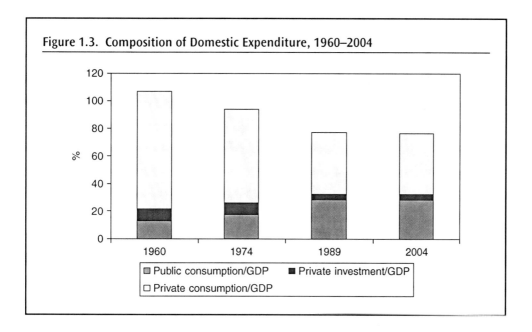

Figure 1.3. Composition of Domestic Expenditure, 1960–2004

major drop of roughly 39 percent in GDP per capita in 1993. In addition, changes in oil prices provoked economic contractions during that period when GDP per capita declined at an average rate of 2 percent per annum. From 1997 to 2004, GDP per capita grew at an average rate of 4.2 percent per annum with the biggest increase observed in 2002 (about 13 percent). In mid-2002 gradualist economic policies were adopted and by 2004 the government managed to bring inflation down and become relatively more transparent in the oil and fiscal sectors. Currently, the level of GDP per capita stands at US$ 1,784, which is still half of the level observed in 1973.

The shift to a command economy right after Independence led to changes in the way resources were allocated in the economy. Before 1974, total government consumption was less than 20 percent of GDP with private consumption at about 70 percent of GDP (see Figure 1.3). After Independence, these proportions changed substantially and today public consumption is about 28 percent of GDP while private consumption accounts for 44 percent of GDP. The shift from a market economy in 1974 to a command economy in 1975 meant in practice a structural change in the way resources were allocated in the economy. One of the important changes was in the economy's price system as the government introduced (i) price controls for a large number of goods and services, (ii) price ceilings for some products, and (iii) guaranteed minimum prices for agricultural and livestock, thus misaligning relative prices and distorting price signaling mechanisms.

In the period of strong centralization, practically all prices of goods and services in the official market were controlled by the Government. The policy of controlled prices was administered by the National Planning Office (*Direcção Nacional de Preços*, now *Direcção Nacional de Preços e Concorrência*). The Ministry of Finance and other sectoral Ministries were also often involved in price control decisions and the responsibility for monitoring and supervising controlled prices belonged to the Office for Inspection of Economic Activities (*Direcção Nacional de Inspecção e Investigação das Actividades Económicas*), a Department

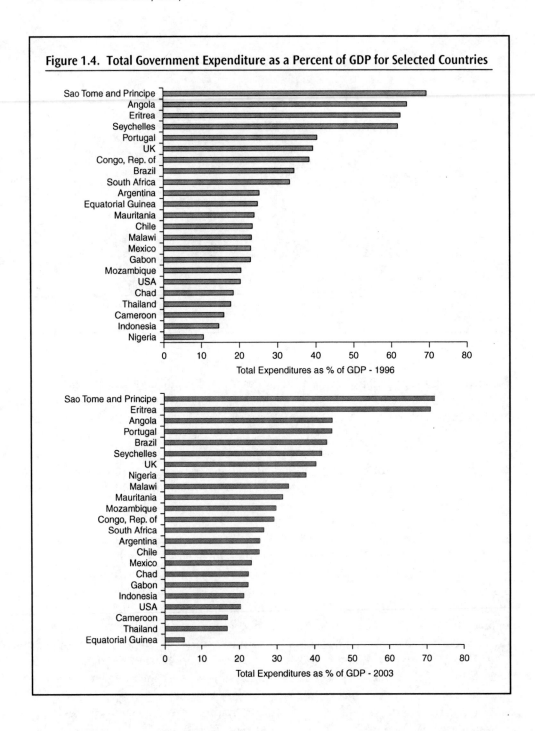

Figure 1.4. Total Government Expenditure as a Percent of GDP for Selected Countries

of the Ministry of the Interior. The artificial stability imposed by price controls and the rapid increases in cash balances held by consumers led to large and increasing gaps between the supply and demand of practically all goods and services in the official market.

With the option for centralization, the importance of the public sector in the Angolan economy grew and is now considered high by international standards. Before independence, government expenditures as a share of GDP averaged less than 15 percent at current international prices. Between 1974 and 1975, the ratio augmented to 25 percent of GDP and since then has remained amongst the highest in Africa and elsewhere (see Figure 2.3). The highest share was registered in 1999 reflecting the scaling-up of the government's war efforts against the armed opposition movement UNITA. With the end of the conflict in 2002, this ratio has been declining steadily, reaching 37.5 percent of GDP in 2004, which is still high by international standards. The large size of the public sector has led to monetization of fiscal deficits in the past which contributed to high inflation, macroeconomic instability, and vulnerability to external shocks.

The next chapter assesses progress towards macroeconomic stability in a time of transitions. As discussed in this opening chapter, the prolonged war, the rapid development of the oil sector and the policies pursued after Independence, have left the Angolan economy in a unique situation, characterized by very uneven indicators of development. For example, Angola's substantial oil production leads to a per capita GDP (approximately US$1,800 in 2004) that would place it among lower middle income countries. The chapter also highlighted that the transition to a market economy was initiated in the mid-1980s and has yet to be fully completed. The next chapter explains what has failed and what is working well in the search for macroeconomic stability and discusses the likely tensions that the government will have to face and manage to complete the transition not only to a market economy but to a viable democracy that can secure sustainable development. These tensions include the need to maintain fiscal discipline in a context of rapidly increasing oil revenues, the development of the interior in tandem with the cities, dealing with Dutch Disease-related phenomena (such as exchange rate appreciation, corruption and waste of public funds), and consolidating the transition to a multi-party democracy.

CHAPTER 2

Macroeconomic Performance in a Time of Transitions

After Independence, Angola embarked on a system of centralized economic and political management that only in the mid-1980s started to be reviewed. The transition to a market economy took impetus with an ambitious reform program introduced in 1987 that aimed at stabilizing the economy, securing fiscal discipline, encouraging the development of the private sector, and abandoning the system of administered prices. Progress on this agenda has been sluggish and only after the early 2000s, aided by the fortuitous role played by growing oil revenues, the government succeeded in curbing inflation and achieving an incipient macroeconomic stability. In the face of a favorable external outlook, the country now has the opportunity to consolidate not only the transition to a market economy but also the transition to a multi-party democracy. This chapter looks into the progress in terms of macroeconomic performance and outlines some of the tensions that will need to be addressed to secure the recently achieved macroeconomic stabilization and the consolidation of a viable democracy—both of which are of key importance for the overall objective of a peaceful reconstruction.

The Transition to a Market Economy

A transition to a market economy was initiated in the mid-1980s with an ambitious program of reforms. As described in Chapter 1, the Angolan economy functioned under Marxist principles and a centralized planning system after Independence. The transition to a market economy initiated in the mid-1980s when the government adopted in 1987 an ambitious program of Economic and Financial Restructuring, designated as SEF (*Programa de Saneamento Económico e Financeiro*), and oriented towards the following two main objectives:

1. Stabilization of the financial situation, by reducing internal and external disequilibria, which were reflected in inflationary pressures, large budgetary deficits, excessive losses and indebtedness of many enterprises, serious deterioration of the

financial situation of the banking system, accumulation of arrears in foreign payments, and difficulties in servicing the external debt;
2. Reform of the economic system, in order to increase productivity, improve the allocation of resources and create the conditions for a faster rate of economic growth and equitable development in the future.

The reforms ranged from budgetary discipline to rescheduling of the external debt and adjustments to the planning system. At the beginning of 1989, the authorities approved a "Program of Economic Recovery" (*Programa de Recuperação Económica*—PRE) oriented to the two main objectives of starting the process of macroeconomic adjustment and of promoting the rapid recovery of production. The PRE initiated the implementation of the economic reforms announced in the SEF, which included the following: (1) the reduction of the budget deficit of the state budget; (2) the adoption of new solutions to finance the budget deficit; (3) the restructuring of the financial situation of public enterprises; (4) the reform of domestic credit policies; (5) the rescheduling of external debts; (6) adjustments of controlled prices; and (7) adjustments in the exchange rate.

On the structural side, the reforms aimed at increasing the role of the private sector and at gradually reducing the importance of the state in the economy. In what is concerned with structural reforms to increase the efficiency of the productive system, the SEF envisaged a more important role for the private sector, more autonomy for public enterprises, a revision of the law on foreign investment and improvements in the planning system. The SEF explicitly admitted that smaller public enterprises should be transferred to the private sector and that state ownership should remain concentrated largely in key enterprises with strategic roles. As regards improvements in the planning system, the SEF envisaged to achieve better coordination between the Annual Plans, the State Budget, and the Foreign Exchange Budget, and more decentralization of planning activities from the Planning Ministry to the planning organizations at regional levels.

Despite the appropriate focus, the reforms did not yield the expected results. Government efforts to implement the SEF and the PRE proved unsuccessful and between 1989 and 2000 some 12 different macroeconomic stabilization programs were introduced with equally frustrating results. On average, there were 1.2 programs per year and each of the programs lasted for a period of 10.6 months (see Table 2.1). The main obstacles to the lasting and effective stabilization of the economy continued to be throughout this period the lack of fiscal discipline, the excessive centralization in the planning system and the resulting bureaucratization of the economy, and the inefficiency of the state in promoting the growth of productivity. Fiscal deficits remained high during the 1990s making it difficult for the authorities to reduce inflation, the oil economy remained as the main source of revenues to the state without productive links with the other sectors of the economy, and the priorities of the war continued to condition government expenditures which focused primarily on consumption and military expenditures and neglected social and development spending (notably on health, education, and infrastructure).

Macroeconomic Instability and Dollarization in the 1990s

In the context of failed stabilization attempts, Angola struggled for years with stop-go cycles of inflation and devaluation coupled with extremely low confidence in the domestic

Table 2.1. Macroeconomic Stabilization Programs Adopted between 1989 and 2000

Year of Adoption	Name of the Program	Duration in Months
1989–1990	PRE–Programa de Recuperação Económica Program of Economic Recovery	14
1990 (May)	PAG–Programa de Acção do Governo Program of Government Action	8
1991	PN–Plano Nacional–National Plan	12
1992	PN–Plano Nacional–National Plan	7
1993	PEE–Programa de Estabilização Económica Program of Economic Stabilization	3
1993 (March)	PEG–Programa de Emergência do Governo Program of Emergency of the Government	8
1994	PES–Programa Económico e Social Economic and Social Program	12
1995–96	PES–Programa Económico e Social	18
1996 (June)	Programa Nova Vida–New Life Program	6
1997	PES–Programa Económico e Social Economic and Social Program	12
1998–2000	PERE–Programa de Estabilização e Recuperação Económica Program of Stabilization and Economic Recovery	12
1999–2000	Estratégia Global para a Saída da Crise Global Strategy to Exit the Crisis	15

Source: Alves da Rocha (2001), p. 64.

currency. As discussed above, a succession of failed or only partially successful stabilization plans marked the country's economic policy history since its first major attempts to stabilize in 1987. During the period up to mid-1996, there were successive inflationary peaks that would be temporarily brought down by the adoption of a plan, only to be followed by an even higher peak after some months (see Figure 2.1). In 1997, there was a "structural break" on the inflation rate time series as the Government introduced the so-called "Nova Vida" Plan. Up until then, inflation was not only four digits, but also very volatile, due to continuous adoption of new (and unsuccessful) plans. With the "Nova Vida" Plan, and the introduction of a fixed exchange rate (to the dollar), inflation was halted for a while, but crept up again in 1997 as it became clear that fiscal adjustment was again being postponed. However, since the introduction of the "Nova Vida" Plan, the connection between dollarization (as measured by the share of foreign currency deposits to M2) and inflation became much closer.

Responding to macroeconomic instability, the phenomenon of dollarization appeared as a result of financial adaptation in the economy. To a certain extent, dollarization can

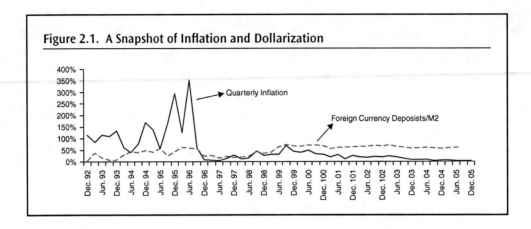

Figure 2.1. A Snapshot of Inflation and Dollarization

be seen as an endogenous process triggered by macroeconomic instability, high inflation, and the resulting lack of confidence in the domestic currency. Countries that display these characteristics can either facilitate financial adaptation when they allow residents to hold financial assets indexed to a foreign currency or to some other stable unit of account, or they can stifle the adaptation when they impose additional distortions that lead to financial disintermediation and capital flight. These are some of the options governments have at their disposal to try to minimize the adverse effects of macroeconomic instability. They are all second-best, and they all entail costs. Theory and evidence suggests that the latter option is probably the most costly and, in this sense, Angola is on the right track.

More recently, the government's economic policy has yielded positive results, but sustainability will demand further reforms. As discussed in the following section, with the implementation of a more rigorous monetary policy, the restriction of monetary financing of the budget deficit since 2002, and the implementation of an active exchange rate policy since September 2003, inflation has been significantly reduced. However, the outlook is subject to significant risks, which must be addressed by government actions. Most importantly, in an uncertain environment for oil production and prices, public expenditure growth needs to be set in a medium-term context to avoid the boom and bust cycles that have undermined stability and development in other oil-producing countries. The following paragraphs highlight progress obtained so far and the tensions that will need to be managed to complete the transition to a market economy and to a viable democracy.

More Revenues and Less Inflation

The economic outlook in Angola has been transformed by the peace agreement of 2002 and by positive developments in the oil sector. With the end of violent conflict and the return of more than 4 million IDP to their original communities since 2002, agricultural production has picked up and the non-oil economy has shown signs of a vigorous recovery in Angola. Although official and detailed data on the non-mineral economy is scant, the lively and vibrant informal economy that is now seen in the streets of Luanda is a visible leading indicator of strong economic performance. There have also been encouraging signs of recovery in public services, construction, and infrastructure rehabilitation. Oil production, which currently accounts for 55 percent of GDP is expected to double to 2 million barrels per day by 2007. Largely as a result of increasing oil production combined with rising international oil prices, real GDP is estimated to have grown by 20 percent in 2005, while the economy outside the mineral sectors is estimated to have grown at an annual rate of roughly 10 percent over the last 3 years. Current projections indicate that GDP is expected to grow by 15 percent in real terms in 2006 and by 30 percent in 2007, one of the highest growth rates in the world.

The macroeconomic framework for 2006 and 2007 is highly favorable and accommodates significant changes on the revenue and spending sides. Table 2.2 presents a hypothetical scenario for the next two years on the basis of preliminary assumptions about economic growth, oil production, evolution of world oil prices, and revenue capability. The figures are based on information as of March 2006 and reflect the macroeconomic framework agreed between the authorities and the Fund during the 2006 Article IV consultations. In our estimates, total government revenues are expected to remain at a level close to 38 percent of GDP until 2007. On the expenditure side, spending is estimated to decline from 38.5 percent of GDP in 2004 to 35.7 percent of GDP in 2005. In the pursuit of long-term fiscal sustainability, spending should gradually decline in 2006 and 2007 as a share of GDP.[16] Such gradual decline in public spending as a share of GDP is not politically unrealistic insofar as real GDP is estimated to have grown by 20 percent in 2005 and to grow by an average 24 percent in 2006–2007 supported by strong performance in the oil sector and steady recovery of the non-oil economy.

There have been commendable successes towards macroeconomic stabilization, but there should be a stronger emphasis on the continuing deficiencies in policy design and implementation. Figure 2.2 shows progress on a number of macroeconomic variables since the year 2000, including oil production. The stabilization obtained so far, however, needs to be strengthened with improved coordination of the fiscal policy with monetary and exchange rate policies. These policies need to spell out a consistent strategy to absorb the upcoming oil windfall without inhibiting growth outside the mineral sectors. To avoid the boom and bust that have undermined stability and development in some other oil-producing countries, new public spending in the future should be set in a medium-term context. In addition, the authorities should consider the adoption of a monetary anchor, with the responsibilities for executing monetary policy defined for the Central Bank in order to guarantee a downward trend in inflation even in the face of an external shock. These issues are discussed in depth in the next Chapter.

---

16. In the draft 2006 budget recently finalized by the Government, the authorities are projecting a fiscal deficit of 6.9% of GDP in 2006 and an annual inflation rate of 10%. The fiscal numbers in our macroeconomic framework are different from those presented by the Government in the 2006 budget because our estimates use higher oil prices for 2006 ($56/barrel) than those used by the Government in their 2006 budget ($45/barrel).

Table 2.2. Macroeconomic Framework, 2003–2007

	2003	2004	2005 Est.	2006 proj.	2007 proj.
	(Percentage change, except where indicated)				
National income and prices					
Real GDP	3.3	11.2	20.6	14.6	30.2
Oil sector	−2.2	13.1	26.0	15.0	40.9
Non-oil sector	10.3	9.0	14.1	13.8	13.7
GDP per capita (in U.S. dollars)	959	1,322	2,129	2,780	3,614
GNDI per capita (in U.S. dollars)	848	1,157	1,866	2,449	3,082
Consumer price index (annual average)	98	44	23	13	8
Consumer price index (end of period)	77	31	19	10	7
Money and credit (end of period)					
Net domestic assets[2]	12	−9.7	−9	−102	−52
Broad money[2]	67	50	60	43	29
M2 velocity (non-oil GDP/average M2)	3.35	3.55	3.33	2.92	2.65
Base Money in real terms (percent change)	−0.5	19.1	40.2	30.0	16.0
	(Percentage of GDP, except where indicated)				
Fiscal accounts					
Total revenue	37.9	36.9	38.0	38.0	37.9
Of which: oil	27.9	28.4	30.1	30.0	30.5
grants	0.8	0.5	0.2	0.3	0.2
Total expenditures	44.3	38.5	31.2	35.7	32.5
Overall balance (accrual basis)	−6.4	−1.6	6.8	2.2	5.4
Non-oil fiscal balance (accrual basis)	−35.1	−30.4	−23.6	−28.0	−25.3
Overall balance (cash basis)	−5.6	−3.7	6.0	1.6	3.1

External sector					
Current account balance (including transfers; deficit −)	−5.1	3.5	12.9	8.8	12.4
External debt (in billions of U.S. dollars)	10.2	10.8	12.6	15.0	16.3
External debt-to-GDP ratio	73.1	54.5	38.5	34.1	27.7
Debt service-to-net-export ratio[3]	16.5	16.4	10.5	4.8	6.4
	(In millions of U.S. dollars, except where indicated)				
Net international reserves (end of period)[4]	790	2,023	4,140	9,252	13,920
Gross international reserves (end of period)[4]	800	2,034	4,147	9,261	13,927
Memorandum items:					
Gross domestic product (in millions of US$)	13,956	19,800	32,810	44,103	59,019
Official exchange rate (kwanzas per U.S. dollar; end-of-period)	79.1	85.6	80.8	—	—
Gross domestic product (in billions of kwanzas)	1,041	1,652	2,860	3,539	4,839
Oil production (thousands of barrels per day)	875	989	1,247	1,434	2,019
Price of Angola's oil (U.S. dollars per barrel)	28.2	36.4	50.1	56.6	57.4
Non-oil fiscal balance/GNDI	−38.8	−34.3	−26.6	−31.5	−29.4

Sources: Angolan authorities and IMF and World Bank staff estimates and projections.
[1] End of period. A positive sign denotes appreciation.
[2] As percentage of beginning-of-period M3.
[3] In % of exports net of oil-related expenses such as oil-related imp. of goods and services and oil companies' remittances.
[4] Includes government deposits in overseas accounts.

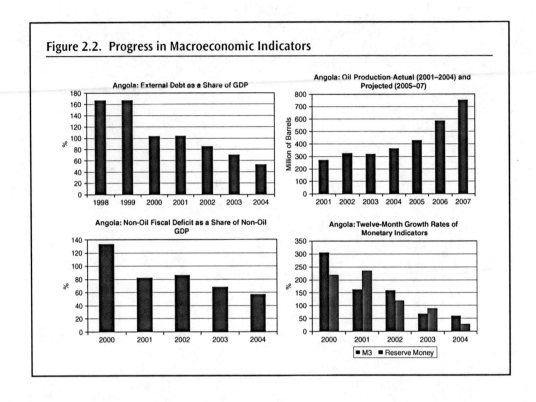

Figure 2.2. Progress in Macroeconomic Indicators

Bringing Inflation Down with Success

The root cause of past inflationary episodes in Angola was the monetization of its fiscal deficits. Angola's main source of fiscal revenue is through the taxation of the oil sector, including the state-owned oil company Sonangol. As a result, fiscal revenues have been excessively vulnerable to international crude oil price volatility and have not always been able to keep pace with expenditures. The insufficient control of public spending, including notably large extra budgetary expenditures and the sizeable operational deficit of BNA, have induced large increases in base money. Additionally, in the past, favored interest groups, including Sonangol, have used arbitrage and other tactics to benefit from high inflation, for example, by delaying payments in domestic currency for oil and other sales received in hard currency.[17] Until 2002, this combination of affairs had actually created positive incentives for high inflation.[18]

---

17. A detailed discussion of public finance management issues can be found in the Bank's PEMFAR report, published in February 2005 (see World Bank 2005).

18. From a political economy point of view, a centralized economic system that functions based on controlling markets encourages the development of a wealthy elite which tends to create mechanisms to guarantee the appropriation of profits irrespective of exchange rate and price swings so they are largely indifferent to macroeconomic shocks and the need to stabilize the economy. In fact, the wealthy can lose from economic reform to the extent that competitive markets and transparent public finances shrink the scope for rent extraction. Some commentators argue that this was actually one of the reasons beyond the war situation that could be used to explain why reform stalled through the 1990s (Aguilar 2001).

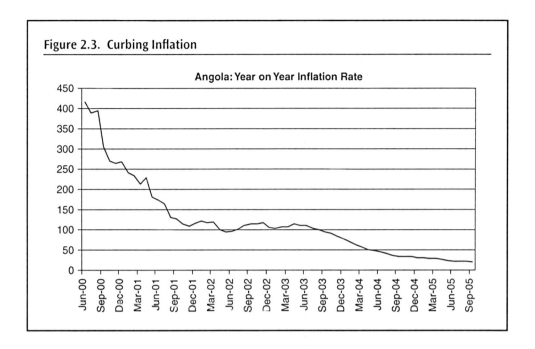

Figure 2.3. Curbing Inflation

More recently, government's efforts to reduce inflation have been successful. Between 1999 and the peace agreement of 2002, annual consumer price inflation fell from around 300 percent to around 100 percent (see Figure 2.3). Following the adoption of a stabilization program in September 2003, inflation fell sharply again and by December 2004 the 12-month inflation rate had declined to 31 percent. The improvement was largely due to the government's avoidance of money creation for deficit finance purposes together with smaller fiscal deficits in 2003 and 2004 (that dropped from 6.5 percent of GDP in 2003 to 1.5 percent of GDP in 2004, on a commitment basis) and an estimated fiscal surplus of 6.8 percent of GDP in 2005. Since 2003, government spending has been increasingly funded with resources obtained through direct sales of foreign exchange in excess of $2 billion in 2003 and 2004, respectively, which has increased Angola's external liabilities.[19] The non-oil fiscal deficit as a share of non-oil GDP has also declined substantially since 2000 from around 130 percent to close to 63 percent in 2005. In 2005, the cumulative rate of inflation dropped to 18.5 percent and the projection for 2006 is of an annual rate of 10 percent.

Monetary aggregates have been kept under control contributing to lower inflation. In the past, the Central Bank of Angola traditionally accommodated unplanned, extra-budgetary expenditures and followed a practice of monetizing the fiscal deficit. Such policy was responsible for rapid money supply growth and high inflation and has been discontinued

---

19. A detailed description of the kinds and magnitudes of intervention in the foreign exchange market in Angola is available in the *2004 Angola Economic Report* published by the Center of Studies and Scientific Investigation of the Catholic University of Angola—see CEIC (2004).

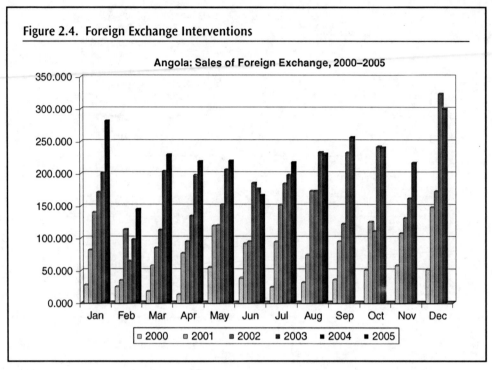

Source: National Bank of Angola.

in recent years, and this is a welcome development. Since 2003, reduced reliance on the domestic banking sector to fund the public sector has contributed to a considerable fall in the growth rate of the monetary aggregates. In the 2004 Article IV consultation, the IMF reported a substantial fall in reserve money growth from 160 percent in the 12-month period prior to September 2003 (the beginning of the Government's so-called "hard kwanza" stabilization policy) to 26 percent one year later, while the growth in broad money fell from 125 percent to 50 percent. In 2005, the situation remained encouraging with reserve money growing by less than 30 percent up to June while broad money expanded by 33 percent in the 12-month period up to September. The slow growth in reserve money can be attributed to net foreign currency intervention in this period, which exceeded US$1.2 billion, using proceeds from external borrowing and higher sales of government securities.[20]

There are clearly beneficial consequences associated with the current policy to combat inflation. First, removal of excess liquidity from circulation reduces the inflationary pressures

---

20. Additionally, the IMF reports strengthening of domestic open market operations from July 2003 through the introduction of treasury bills with maturities of between 28 and 182 days. Longer-term bonds (2 to 7 years) were also issued to clear arrears to domestic suppliers. With the deepening of the domestic debt market, commercial banks developed new instruments, including repurchase agreements and bankers' acceptances whose holdings are included in broad money (M3) which is dominated by deposits in foreign currency.

Table 2.3. International Experiences on Macroeconomic Stabilization Programs

Country	Stabilization Episode	Exchange Rate Arrangement	Initial Inflation Rate (%)	Lowest Inflation Achieved (%)	Was the Stabilization Sustainable?
Israel	Jun 85–Sept 86	Fixed, Crawling Peg	1128.9	50.1	Yes
Brazil	Feb 86–Nov 86	Fixed	286.0	76.2	No
Mexico	Dec 87–Dec 94	Fixed, Crawling Peg	159.0	6.7	No
Peru	Aug 90–Dec 99	Monetary anchor/ dirty float	12378.0	10.2	Yes
Dominican Rep.	Aug 90–Dec 99	ER unification and floating	60.0	2.5	Yes
Argentina	Apr 91–Dec 01	Currency Board	267.0	−0.3	No

Source: Gasha and Pastor (2004).

deriving from money expansion permitting a decline in the rate of inflation. Second, the use of foreign exchange for the purpose of mopping up liquidity contributes to stabilize the exchange rate. Third, keeping the exchange rate stable implies in a corresponding constancy in the prices of imported goods, eliminating inflationary pressures from this source. Finally, avoiding a policy that requires an immediate fiscal adjustment in favor of one which yields price stability while postponing cuts in expenditures generates an immediate and visible success in economic management. These effects combined may help the government to build the necessary political capital for the future, when expenditures will invariably have to be cut down.

The continuity of such policy, however, must be weighed against the costs of keeping it unchanged. The authorities have managed to bring inflation down in Angola without having to promote an upfront fiscal adjustment, but the international experience reveals that it is questionable whether this kind of policy is sustainable without stronger efforts towards a balanced fiscal budget. The main factor that distinguishes success from failure in a sample of countries which have adopted similar price stabilization attempts is the extent to which governments pursued fiscal contraction in support of their stabilization efforts (see Table 2.3). Countries that did not reduce expenditures, as for example Argentina, Brazil, and Mexico, ended their stabilization attempts with renewed inflation, low confidence in the local currency, and severe drawbacks to investment and growth.[21] At the current juncture, in which massive oil windfall revenues are likely to accumulate in a relatively short period of time (see next Chapter), the authorities should consider the

---

21. Calvo and Vegh (1999) argue that lasting stabilization programs are characterized by a significant fiscal adjustment, independently of the monetary arrangements. Agénor and Montiel (1999) review a variety of stabilization attempts in several countries and point to the need to a permanent fiscal adjustment in any serious disinflation program.

adoption of a policy mix that allows for a conservative increase in government spending without sacrificing fiscal discipline. This would require complementing the current anti-inflation policy with the setting up of a medium-term fiscal framework that could yield a balanced budget and that could incorporate the spending needs to repay the peace dividend to the Angolan people in line with the expected increase in oil revenues. The costs of doing otherwise could be too high to bear.

Achieving sustainable inflationary stability is also essential to harness the growth of the non-oil economy. It is a well established fact that the inflation component of market-oriented reform policies should be expected to be growth-enhancing. In the case of Angola, this is particularly relevant for the non-oil economy, given the insulation of the oil-economy due to its enclave nature. This is so because high rates of inflation can be expected to reduce economic growth through a variety of mechanisms which can influence both the rate of capital accumulation and the rate of growth of total factor productivity. One of such mechanisms is that high inflation means unstable inflation and volatile relative prices, which reduces the information content of price signals and distorts the efficiency of resource allocation thus harming the growth of total factor productivity over extended periods. Furthermore, governments that tolerate high inflation have lost macroeconomic control, and this circumstance is likely to deter domestic investment in physical capital.[22]

Beyond the fiscal sphere, there are important concerns associated with the virtual stabilization of the nominal exchange rate that the current policy has generated. As noted above, exchange rate stability may be seen as a beneficial consequence of this policy by working as a major factor limiting price increases of tradable goods. However, the growth of the monetary aggregates in Angola, while slowing, has been faster than would be consistent with the achievement of the inflation objective, with the consequence that the real exchange rate has appreciated significantly. The implication of this is that the policy has been much less effective in reducing the inflation of the non-tradables (see Figure 2.5). In addition, the scaling up of public spending is likely to exert pressure on the domestic price level. To counter that effect and keep the declining trend in inflation, the authorities should resort to both sales of foreign currency and new issues of government bonds to help mop up excess liquidity. In this context, an appreciation of the nominal exchange rate should be not be resisted as it will contribute to reduce inflation.

The External Sector: Oil is Well that Ends Well

The external outlook is positive as a result of rising international oil prices and domestic oil production. The external accounts have moved into a significant surplus in 2005 (12.9 percent of GDP) after a widening of the current account deficit in 2003 (5.1 percent of GDP) which was attributed to higher imports of goods and services related to investments in the oil sector.

---

22. Barro (1997) presents cross-country evidence on the negative relationship between inflation and growth for a sample of 100 countries with annual observations on macroeconomic data for the period 1960–90. His central finding is that, other things equal, a 10% increase in the rate of inflation reduces long-run growth by about 0.025% per year.

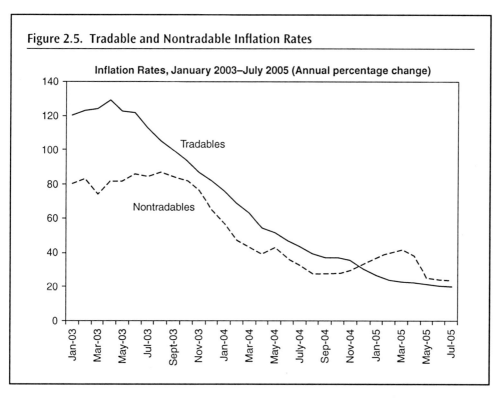

Figure 2.5. Tradable and Nontradable Inflation Rates

Despite substantial interventions in the foreign exchange market, the monetary authorities were able to increase its gross official reserves from US$400 million in 2002 to US$4.1 billion in 2005, which is sufficient to pay for 6 months of non-oil imports (see Figure 2.6). The authorities estimate that in 2006 the level of their net international reserves will

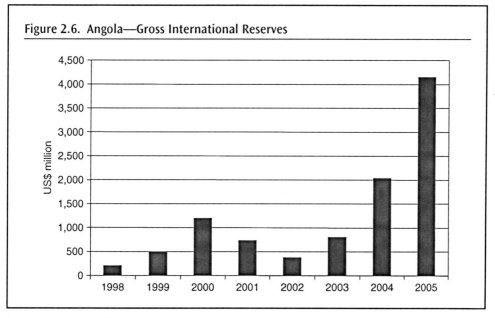

Figure 2.6. Angola—Gross International Reserves

Source: National Bank of Angola.

increase to some US$9 billion level. At the same time, Angola's external debt as a share of GDP has been declining since the late 1990s because of a strong growth rate in the economy's oil production value. In monetary terms, the country's external debt increased from US$8.7 billion in 2002 (more than 60 percent of GDP) to US$12.6 billion in 2005 (38.5 percent of GDP). The external debt-over-GDP ratio is expected to decline to 34 percent in 2006 given the expected rapid growth in GDP.

The Tensions of Transition

Angola still has to overcome tensions that are inherent to a country which is in transition. Entrenched interests undermine structural reforms in several areas which are considered key to a stronger involvement of the international community in the reconstruction process (OECD, 2005). This tension between the government and the IFIs actually reflects an inner issue in Angola that is rooted in a non-transparent conflict inside the ruling MPLA between the so-called reformers, who seem to favor a stronger market-oriented approach in the management of the economy, and the so-called non-reformers, who seem to place a higher value to the role of the state in paving the way for the country's economic development. In order to embark on a long-term sustainable development path, Angola will need to manage these differing views and other tensions associated with conflicting interests arising between the coast versus the interior of the country, rural areas versus urban areas, Dutch-disease related phenomena (notably real exchange rate appreciation), and the normal tensions that accompany the transition from a centralized political system to a muti-party democracy.

Maintaining Fiscal Discipline

Although there may be ample room for public expenditure increases in Angola, capacity constraints and concerns associated with the paradox of plenty call for prudence in defining the public expenditure envelope for the next years. While the consensus in terms of the role of the state in the economy is not settled, policymakers will continue to face the difficulties associated with maintaining macroeconomic stability in the presence of large, petroleum-related foreign exchange inflows. The main problems resulting from a large and volatile oil windfall are discussed in detail in the next chapter but are typically manifested in excessive real appreciation of the exchange rate and a structural shift toward non-tradable sectors. In the context of low absorptive capacity, rapid spending of the oil windfall may also lead to a "waste" of petroleum wealth through unproductive public expenditure. There is always the risk that rapid spending may lead to a rent-seeking behavior to garner the windfall, and this may occur in an environment that is weak institutionally and is thus less able to filter out all of the requests.[23] All of these issues are relevant for Angola and a solution to these problems will require an accelerated pace to structural reforms in the areas of governance, private sector development, and human capital formation, without which

---

23. For example, McMahon (1997), cited in Sarraf and Jiwanji (2001), provides examples of excessive investment in the under-developed non-traded sector during resource booms, due in part to political pressure to prop up ailing industries.

larger public spending may not yield the expected outcomes.

The solutions while straightforward in principle are politically complicated. Ample empirical evidence has shown that natural resource dependence is particularly problematic, since it is easily captured by the ruling elite, removing the incentive for the government to actively engage its citizenry. This destroys both the capacity and the legitimacy of the state, exacerbating social divisions and even leading to direct conflict over the resource itself. Research at the World Bank and elsewhere points to resource dependence as one of the most important causes of civil wars (see Figure 2.7).[24] While an increase in oil revenues increases rent-seeking behavior and

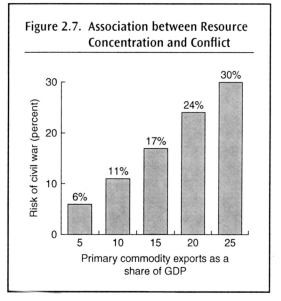

Figure 2.7. Association between Resource Concentration and Conflict

Source: Bannon and Collier (2003).

gives governments an excuse to delay reform, it also allows reform-minded governments to implement changes. When oil prices are high, resource constraints get relaxed, allowing governments to pay off debt or invest in infrastructure and the social sectors—provided fiscal institutions and the commitment to reform are strong. Angola has now a real opportunity to use its oil wealth to mitigate tensions and put the economy on a path of sustainable development.

Developing the Interior and the Cities in Tandem

Diversification of the economy outside the mineral sectors and equitable development will require the reconnection of the interior of the country to the coastal areas. In any part of the world, including Angola, those who control coastal areas and/or ports exercise a degree of control over interior landlocked areas by virtue of their ability to control or charge for access to outside markets. There is a real economic foundation for this: whoever controls access to outside markets can charge a price for the privilege.[25] The same logic applies to landlocked areas within a country, particularly when the country in question is composed of provinces which have some degree of de facto autonomy, or the central government apparatus uses control of ports to generate revenue for its own use. This is no different in Angola than it is elsewhere, with the major ports of Luanda and Lobito/Benguela playing

---

24. Violent secessionist movements can often be traced to oil. Examples include Aceh in Indonesia, Biafra in Nigeria, conflicts in Sudan, Chad, Congo Brazzaville, and in Angola itself (see Busby et. al., 2002).

25. The importance of free access to a port has been shown empirically by Bloom and Sachs (1998) who analyzed the growth records of Sub-Saharan African countries and found that geographical access to a port was a significant determinant of growth prospects.

the most important role, but also to some extent the lesser ports of Namibe, Porto Amboim, Ambriz and others. In colonial times as well as more recently, the central government (be it Portuguese or Angolan) has explicitly and implicitly taxed both agricultural exports (particularly coffee through the government marketing board) and imports through tariffs and import quotas. In order to diversify the economy outside the mineral sectors, the authorities need to create incentives for domestic production enhancement (such as incentives for irrigation, demining, agricultural research, and the strengthening of smallholder agriculture) and invest in infrastructure to reconnect the interior to the coast and thus make the main markets accessible to the poorest.

Infrastructure reconstruction in the cities will need to be done in tandem with the strengthening of the agricultural sector. Even in countries where the alignment of coast with cities/industries and interior with agriculture does not exist one can observe a similar type of tension between agricultural interests versus urban/industrial ones by virtue of the fact that what one party buys, the other one sells and vice versa. In Angola, the coast is not only a gatekeeper, but it is also the location of the major urban/industrial centers, particularly Luanda, while the interior is also the location of the most productive agricultural areas, particularly the central planalto where the principal grain supplies of the country are (or could potentially be) grown.[26] The observation that the principal geographical location of agricultural comparative advantage lies in the interior has a very important corollary: in order to maximize the return to this comparative advantage it will be necessary to invest in this area. These investment needs have the potential to be contentious since at first blush the money would be going to one area of the country seemingly at the expense of other zones. However, the experience of many other countries has demonstrated that to attempt to pursue industrialization without investing in increased agricultural productivity is a self limiting strategy. A possible way of preempting such type of conflict could be through the adoption of a spatial approach for economic development that would link coastal areas with promising areas in the interior of the country.[27] The World Bank could help the authorities to identify these areas and develop further a spatial approach for economic development in Angola.

Dealing with Dutch Disease-related Phenomena

The authorities will need to act firmly to preempt rent-seeking. The Angolan authorities need to be aware of and take serious actions to combat a very common problem associated

---

26. It is from this area that pre-independence grain export surpluses were taken, and where most of agricultural GDP originates (see Kyle, 1997, and World Bank, 1994).

27. Mellor (1995) argues that the contribution of agriculture to economic development takes place through (i) a pre-conditioning stage, when improved infrastructure combines with the strengthening of markets to encourage technological change that triggers expanding productivity; and (ii) a follow-up stage where agriculture plays a decisively constructive role in supporting the structural change of the economy. Furthermore, Mellor notes that where a sound macro policy is combined with effective institutions and an improved transportation system (so that crop specialization for trade becomes possible) then the steady diffusion of technology can sustain an agricultural growth rate of 4–6% per annum. Based on data for fourteen Asian countries during 1960–86, Mellor (1995) calculates the agricultural multiplier at 1.5 that implies that agricultural growth can drive GDP growth rates to 7.5% per annum.

with the so-called resource curse that is materialized in the propensity of mineral (and especially oil) dependent economies to develop problems of rent seeking and corruption.[28] The argument is that the existence of oil income results in a scramble for these rents rather than efforts to engage in more productive activities. In addition, these effects can cause institutions to become weak, which will itself have a detrimental effect on growth.[29] In the case of Angola, where already fragile institutions became weaker because of a prolonged civil war, it is extremely important to improve governance to avoid corruption and waste. Specific recommendations on that area are offered in the next two Chapters.

Successful management of mineral income will require improvements in institutions. In what concerns the handling of oil windfall revenues, the case of Nigeria offers a cautionary example. The recent experience of Nigeria is perhaps the most directly relevant to the Angolan case as corruption and institutional weakness have been found to have a more important effect on the economy than has distortion of the real exchange rate. Recent empirical evidence suggests that even though Nigeria invested a large proportion of their windfall, the weakness of their institutions resulted in "bad" investments with very low returns.[30]

The exchange rate is likely to appreciate in real terms and the problems associated with this appreciation will have to be dealt with. Given the size of the windfall revenues to be accumulated over the next few years in Angola, which can reach up to US$266 billion in net present value, the authorities will have very little margin of maneuver to manage the exchange rate, which is expected to appreciate in real terms. An overvalued real exchange rate creates disincentives for domestic production as it reduces the competitiveness of domestic goods *vis à vis* imported ones. In the case of Angola, where there is an urgent need to diversify the economy away from the mineral sectors to create employment and reduce poverty, the authorities will have to deal with this issue diligently. Some of the policy options available involve (i) the creation of a short-term strategy to absorb the windfall, which has to be respected by both the budgetary authorities and the central bank; (ii) an improvement in the coordination between the fiscal policy and the exchange rate policy; and (iii) the adoption of prudent fiscal policies to complement the windfall absorption strategy. These are issues that are discussed in depth in the next Chapter.

Consolidating the Transition to a Multi-Party Democracy

Angola has started a welcome transition to a multi-party democracy. The last Presidential elections in Angola were held in September 1992 with inconclusive results that led to the resurgence of armed conflicts. A government of national unity was formed in April 1997, comprising the ruling MPLA, UNITA and four other opposition parties. Since then, the Government has been structured around a strong presidentialist regime which is supported by the MPLA. In August 2004, a 14-point timetable was established to guide preparations for the next electoral process, which included elections for the National Assembly

---

28. See for example, Mauro (1995), Leite and Weidmann (1999).
29. Isham et al. (2004) find such a relationship in a cross section of countries.
30. See, for example, Sala-i-Martin and Subramanian (2003) on the case of Nigeria.

in 2006 and for the Presidency in 2007. This timetable provided for the preparation and adoption of a new constitution and an electoral law, as well as the appointment of members of the National Electoral Council in 2005. The exact date of the elections will be determined by the President and the candidates must be approved by the Supreme Court. An Interministerial Committee for the preparation of elections was established in January 2005, but so far the dates of the two polls have not yet been publicly announced.

Old ethnic tensions remain as a latent source of concern. UNITA is grouped around the Ovimbundu ethnic group that comprises 37 percent of the population and has traditionally practiced agriculture on the relatively fertile central west planalto, around Huambo. During the civil war this faction used diamond rents to fund its campaign. The end of the civil war saw the collapse of UNITA, which left its coastal rival, the MPLA in control of the Luanda oil enclave within a war-ravaged rural economy. The victorious MPLA's main support is the coastal Mbundu ethnic group, which comprises one-quarter of the Angolan population (FAO/WB 2004). In the 1992 elections, the MPLA carried its coastal heartland provinces and some provinces in which the Mbundu were in the minority (Kyle 2002). In contrast, UNITA carried its planalto heartland provinces but few provinces where the Ovimbundu were in a minority. In the absence of a unifying development plan to integrate the country and develop the interior, tensions between the two main political parties are not likely to decrease without political dissent. The strengthening of democracy in the country should contribute to reduce both ethnic and social tensions that remain latent as post-conflict wounds in Angola.

Some form of political accommodation will be required to maintain a viable democracy. Entirely apart from ideological or geopolitical concerns, the political issues inherent in the transition to a multi-party democracy are fundamental to any long term view of the Angolan political situation. A core issue is the need for maintaining a viable democracy in a situation where the two main parties have such a strong regional base. In the next elections, any victory at the national level by one side or the other must be carefully managed to avoid it being seen by the loser as a "conquest" as much as an electoral loss.[31] In this context, a possible recipe for long term political accommodation in Angola involves some sort of federalism, where a great deal of power is devolved to the provincial level. Although constraints associated with the lack of institutional capacity to implement a fully fledged fiscal decentralization program are non-negligible in Angola, the government should learn with the experience of the subprograms supported by the FAS project and set up a plan to gradually expand its reach and its scope as a way to cede budgetary authority to lower levels of government. This was actually one of the objectives of the SEF program discussed earlier, and which has not yet been institutionalized in Angola.

While there are huge challenges to achieving the objectives of a peaceful and sustainable recovery of the Angolan economy, there are also options available to accomplish them. It is often said that Angola is a resource-rich but policy poor country. As discussed in the very

---

31. Empirical research suggests that, on average, there is an improvement in growth performance once stable democracy is established in a country. Furthermore, growth under democratic regimes tends to be much more stable than under authoritarians regimes. Stronger democratic institutions tend to restrain corruption and this has the effect of stimulating technological change and hence economic growth (see, Rivera-Batiz, 2003, and Shen, 2002).

next Chapter, the Angolans, as a nation, are about to become even richer. But will the oil and diamond wealth be used wisely to heal the wounds of the war, to help rescuing many trapped by poverty and to pave the way to shared and inclusive growth? This report addresses a number of issues that should be the object of thorough reflection by the Angolans in their quest for a peaceful reconstruction anchored on broad-based and equitable economic growth. The following chapters will help to show the way toward the path of prosperity based on the understanding on how the Angolan economy has gotten to where it is today with the hope that this understanding will expose the way forward. After an analysis of the options available regarding the management of the oil windfall, the report takes a closer look at the microeconomic underpinnings of shared growth in Angola by addressing private sector development constraints and opportunities, the potential of the diamond and agricultural sectors in the near future, and how to improve the welfare of the poor.

CHAPTER 3

Oil Wealth: Policy Options to Manage the Windfall

The Angolan economy will experience a massive windfall with a concomitant fiscal gain during the second half of this decade and throughout the next decade. The fiscal gain, however, is a result of the exploration of the country's oil reserves. Long-term sustainability requires that part of the resource rents be reinvested productively, to compensate for this reduction in natural resource capital by the accumulation of other forms of assets. Because oil rents are to a large part concentrated in the public sector, the question of how the oil revenue should be spent and distributed across present and future generations becomes key to any economic development strategy. This chapter deals with this question, discusses governance and transparency issues in the oil sector, estimates optimal levels of consumption and investment under different production and oil price scenarios, and offers policy options for managing the oil windfall revenues.

The Characteristics of the Petroleum Sector

The government became directly involved in oil production in Angola in the 1970s with the creation of a national oil company that is now a major operator and the sole concessionaire for exploration and production. Oil began to be explored in Angola in the mid-1950s in the onshore Kwanza Basin while the main expansion of the upstream (exploration and production) industry occurred in the 1960s when Cabinda Gulf Oil Company (CABOG), now ChevronTexaco, discovered oil offshore the Angolan enclave of Cabinda. In 1973, oil became Angola's principal export. The government became directly involved in oil exploration in the mid-1970s when, in 1976, Sonangol, the national oil company, was created to manage the oil industry on behalf of the government. In 1978, Sonangol acquired a 51 percent interest in Cabinda and in all onshore concessions and was made sole concessionaire for exploration and production, although operatorship remained with the international companies. In the 1990s, Sonangol invited bids for exploration licenses in deep water (over 200 meters), where there have been several major oil finds in recent years, and subsequently licensed a number of ultra-deepwater blocks.

Angola is now a well-established petroleum producer, with an enviable record of exploration success and associated rapid reserve and production growth, and significant remaining petroleum potential. Most of the premier international oil companies have acquired interests in Angola, including ChevronTexaco, ExxonMobil, British Petroleum (BP), Total, Shell, and Agip. New entrants include some smaller companies, such as Devon, CNR, and ROC. Other possible new entrants include the Chinese oil companies. Proven reserves, currently estimated at 8.8 billion barrels, have quadrupled in the 10 years from 1995 to 2005, with the promise of more to come, depending on the level of exploration activity. Angola currently holds 0.75 percent of world reserves, or 1.9 percent of reserves outside the Middle East, and is Sub-Saharan Africa's second largest oil producer after Nigeria currently producing 1.3 MBD, an increase of 70 percent over 2000 production levels. Production is expected to increase by another 90 percent to 2.6 MBD over the next five years. Overall, production will have increased by 225 percent between 2000 and 2010 (see Figures 3.1 and 3.2).

The future potential of oil production in Angola lies in ultradeep offshore waters. Shallow water fields, principally Cabinda, account for over 50 percent of current production. However, these fields are mature and production is set to decline. A production plateau in the range of 1.5 MBD to 2.5 MBD may be maintained for the next 10 years, but the big increase will come from the development of the major discoveries in deepwater and ultra-deepwater. Figure 3.2 shows Angola's production to date and expected future

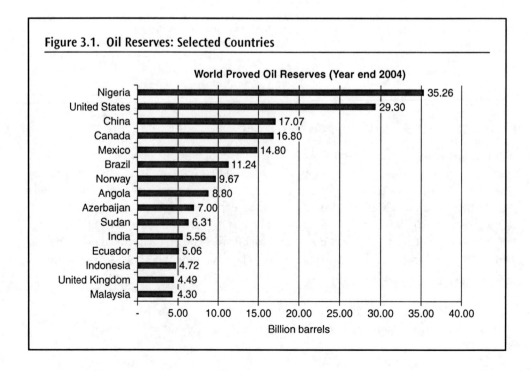

Figure 3.1. Oil Reserves: Selected Countries

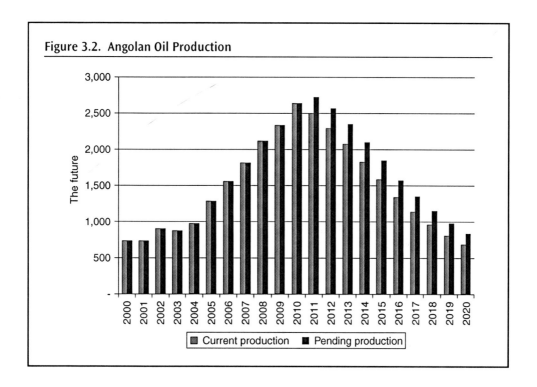

Figure 3.2. Angolan Oil Production

production from shallow water and onshore areas, and deep and ultra-deepwater. The left hand bars in Figure 3.2 shows estimates of future production based solely on currently producing fields and fields committed to development. The right hand bars in Figure 3.2 shows the increased production that might result if major discoveries in the ultra-deepwater Blocks 31 and 32 were to be put into production in the near future. This new production, while substantial, would extend the period of peak production by only two to three years. The shape of the curve beyond this will depend critically on the pace of new license awards or re-awards. Since the average delay from discovery to production is seven years, this suggests that the short-lived albeit dramatic peak shown in Figure 3.2 accurately reflects the medium-term outlook for crude oil production in Angola. If Angola wants to change this picture, new exploration wells should be drilled now.

The pace of new exploration activities may be affected by the costs of new projects in offshore deepwaters. Both capital and operating costs in Angola are relatively low due to the mild environment. For near-shore, shallow water areas, where fields are relatively mature and closely spaced, the majority of capital expenditures are confined to drilling and workovers, rather than more expensive infrastructure. For deepwater blocks, costs are similar to those experienced in comparable projects in other parts of the world such as Nigeria, Brazil, and the Gulf of Mexico. While unit costs may be relatively low, the absolute cost of development projects in Angola, given their typical scale, is large. Appraising and developing the inventory of deepwater discoveries is straining the resources of even the major international companies. Development decisions are complicated by the need to

> **Box 3.1: Petroleum Sector Data**
>
> The principal source of reserve, production, and economic estimates contained in this Chapter is Wood Mackenzie's Global Economic Model (GEM) reports for Angola. Detailed data used to prepare figures and tables are contained in Annex A. Figures are published here with the consent of Wood Mackenzie.
>
> Wood Mackenzie is an established and well-regarded independent petroleum consultancy based in Edinburgh, Scotland. The data and estimates contained in Wood Mackenzie reports are based on semi-annual visits by professional staff to interview industry operators in the country in question.
>
> Angola has engaged another recognized consultancy, AUPEC, from Aberdeen, Scotland, to build a very detailed sector model for its own use. This model, now being updated and nearing completion, will provide important guidance to the Government in tax administration and macroeconomic planning.

smooth the outflow of capital and to separate projects to coincide with the availability of development teams. Sonangol, the State partner of these companies, has been similarly challenged to come up with the resources needed to finance its share of joint venture costs. Sonangol holds a 20 percent equity stake in several deepwater blocks where costs, as Table 3.1 suggests, are significant. Even if the pace of new licensing were not to slow exploration and development activity, these considerations might.

In addition to oil, Angola has substantial natural gas reserves. Current estimates suggest that natural gas reserves in Angola total 1.6 trillion cubic feet (TCF). New discoveries could push Angolan proven reserves to 10 TCF and possibly as high as 25 TCF. Approximately 85 percent of the natural gas produced in Angola is produced in conjunction with oil, and is flared into the atmosphere. The remainder is re-injected into the producing reservoir to aid oil recovery, and/or processed into liquefied petroleum products. ChevronTexaco, Sonangol, and other international petroleum companies in Angola are developing a project to convert

Table 3.1. Unit Cost Comparisons for Selected Countries

		Development Costs $ Per Barrel	Operating Costs $ Per Barrel
Deep Water Exploration Costs			
Angola	Deep	3.43	2.79
Brazil	Deep	2.96	3.71
Nigeria	Deep	2.69	2.56
Norway	Deep	2.64	6.12
USA (Gulf)	Deep	3.14	2.71
Equatorial Guinea	Deep/Shallow	4.13	2.84
Shallow Water Exploration Costs			
Angola	Shallow	4.05	4.60
Brazil	Shallow	2.51	3.45
Congo	Shallow	4.42	3.64
Malaysia	Shallow	1.27	1.47
Nigeria	Shallow	2.70	2.03

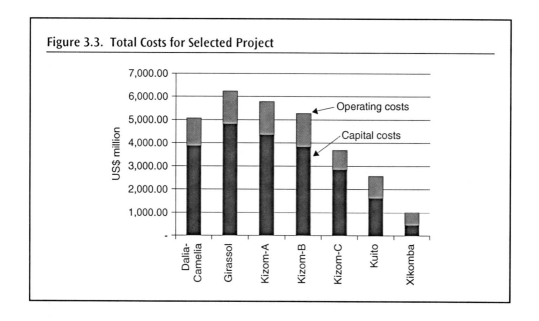

Figure 3.3. Total Costs for Selected Project

natural gas from offshore oil fields into liquefied natural gas (LNG) for international markets. The LNG facility will be located at Soyo and should add more than US$1 billion a year to exports by the end of the decade.

The Legal Framework Governing the Sector

Angola's fiscal terms share many features with petroleum fiscal regimes applied elsewhere in other major petroleum producing countries. Companies operating in Angola do so under two different fiscal regimes—the Concession Agreement with tax and royalty payments, and the PSAs with production sharing (see Box 3.2 on the underlying legal framework). A few comments are nevertheless in order. The terms appear to provide contractors with adequate incentives to undertake a wide-range of projects. The terms also appear to be "competitive," that is, rates of "government take" measured by the net present value of all government revenues as a percentage of pre-take project net present value are comparable to the rates of take found in other major oil producing countries (see Figure 3.4).

The current fiscal regime is complex and its management requires adequate institutional capacity. Within each of the two existing regimes a further layer of complexity exists with different vintages of contract containing different provisions. Further, most of the key elements of PSAs have been negotiable and specific parameters, such as bonus, cost recovery, terms, depreciation, investment uplift, and production sharing scales may differ among contracts within the same vintage. The PIT also differs between concession agreements and PSAs. The cost and complexity of administering such a regime can be very burdensome. A significant level of skilled resources is required to administer the regime effectively. The fact that each PSA is negotiated individually is time consuming and costly. The burden of fiscal administration falls primarily on Sonangol and the regulating ministries—the Ministry of Petroleum and the Ministry of Finance. Measures to ensure adequate institutional capacity

Box 3.2: Legal and Contractual Framework

Petroleum Activities Law. The most important primary legislation governing Angola's petroleum sector is the Petroleum Activities Law. Recently passed, this law supercedes a 1978 version, although many essential provisions remain unchanged. The Law makes clear that the State owns all of the petroleum mineral rights, and provides that the State oil company, Sonangol, is the sole concessionaire with rights to all exploration and production activities. The Law provides that companies wishing to carryout exploration and production activities can do so only in association with Sonangol, and that form of association must be a commercial company, joint venture, or production sharing contract. Sonangol may become an equity participant in a field. Sonangol has an exploration and production company for this purpose (Sonangol P&P) and, as noted above, has exercised its right to equity participation in the development of several deepwater blocks.

Concession Decrees. The mechanism for a block or license award is through a concession decree issued by the Government. These decrees grant concession rights to the concessionaire, Sonangol, and establish the latter's key obligations. These include, *inter alia,* execution of approved work plans and marketing of the Government's share of profit oil (see discussion below under Production Sharing Agreements). Sonangol is the sole concessionaire and is entitled to retain up to 10 % of the proceeds of its marketing of the Government's share of production in order to meet costs incurred in discharging its role as concessionaire.

Sonangol's role as concessionaire has been the source of dispute on at least two counts (a) the perceived conflict of interest in its multiple roles as quasi-regulator (concessionaire), exploration and production company, and supplier (through affiliation) of goods and services to the upstream oil sector; and (b) the expenses received as compensation for its role as concessionaire. In general, however, the concessionaire decrees cover appropriate topics and are clear and modern in style.

Concession Agreements and Production Sharing Agreements. Through Sonangol the Government has entered into two types of agreement with the international oil companies—the Concession Agreement, which applies to onshore and shallow water blocks (principally the Cabinda area), and the Production Sharing Agreement (PSA), which applies to deepwater and ultra-deepwater areas. The two types of agreement are defined primarily by their different fiscal arrangements, which are discussed in the next section.

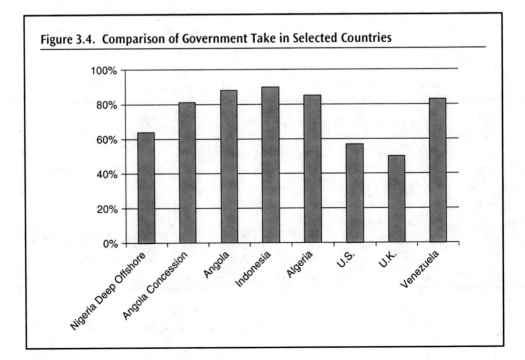

Figure 3.4. Comparison of Government Take in Selected Countries

Table 3.2. Fiscal Terms for Petroleum Exploration Contracts in Angola

Cabinda Concession[32] (from Jan-1982)	Terms of Vintage Contracts (1979 through 1991)	More Recent PSAs (post 1991)
■ 20 percent royalty. ■ A Petroleum Transactions Tax (PTT) charged at a 70% rate on income subject to the Petroleum Income Tax (PIT), except with no deduction for either royalty or interest payment. Two additional items are, however, deductible (a) a petroleum allowance of $4.00 per barrel in 1982, escalated annually at 7%; and (b) an investment allowance, or uplift, or 50% on capital invested. Both of these allowances are deducted once the contractor has achieved a threshold rate of return (23%). ■ PIT at a rate of 65.75% on final profit after deduction of all costs, including royalty, interest, and PIT. ■ Production areas are ring-fenced for tax purposes, i.e., costs incurred with respect to one development cannot be offset against taxable income from another area, with the exception of exploration costs.	■ A negotiable signature bonus. Early payments were relatively modest (less than $10 million). ■ A provision for early recovery of cost or cost oil up to a ceiling of 50% of the value of gross production. Recoverable costs include an uplift on capital expenditures of 33% to 40%. ■ Sharing of remaining production, or profit oil, between the Government and contractor as a function of cumulative production. The negotiable share typically varied between an initial 40% to the Government, increasing to 90% over 100 million barrels of production. ■ A price cap set at $20 per barrel in 1984 and escalated with inflation above which all revenue goes to the Government. ■ PIT at a rate of 50% applied to the contractors profit oil after deduction of the price-cap fee. ■ Ring fencing of fields for cost recovery and PIT purposes, except for exploration costs, which can be consolidated within a block. ■ An uplift on capital expenditures of 40% for cost recovery purposes.	■ The cost recovery allowance has been increased to 55% in particularly deepwater. ■ Signature bonuses vary widely. Early deepwater blocks attracted relatively modest ($20 million) bonuses, while highly prospective deepwater blocks awarded later have attracted bonuses in excess of $200 million. Bonuses are neither cost recoverable nor deductible for PIT purposes. ■ Profit oil is shared between the Government and contractor as a function of the contractor's achieved after-tax rate of return. In a typical PSA, the Government share increased in tranches from 25% at a contractor return of less than 15%, to 90% where the contractor return exceeds 40%.

in this area deserve to be assigned very high priority. With an estimated $15 billion in fiscal revenues due in 2005 alone, effective fiscal administration is critical.

The structure of fiscal terms appears to create incentives in favor of contractors, but it also implies in the concentration of cost recovery, which can result in lower annual rates of government take over time. The structure of fiscal terms described in Table 3.2 implies that the

---

32. Slightly different terms apply to different areas within the concession, and to the onshore Congo and Kwanze concessions.

production allowances, investment uplifts, and generous cost recovery provisions, will tend to "front-load" cash flow in favor of the contractor. This is desirable in so far as it provides an incentive to the contractor to invest by reducing perceived risks and protecting marginal projects against the very high marginal rates of take that apply. At the same time, such terms create a need to manage expectations in those situations where a "bunching" of cost recovery in the sector may result in annual rates of government take that appear low relative to when full cycle calculations of take would show. When production sharing is based on cumulative production, as it is for the early PSAs, this can create disincentives to investment in often important rehabilitation or secondary recovery projects late in a field's life. Under the cumulative production formula, the contractor may at this point have passed the highest production threshold and be facing the highest rate of government take. Furthermore, the exclusion of exploration costs from fiscal ring fencing provisions is attractive as an incentive to explore, but it limits that incentive (early tax write-off of exploration expenses) to those companies that are already "players" in the sector, namely those with already established production income.

The Organization of Fiscal Administration

Current roles in fiscal administration are shared by the Ministry of Finance, the Ministry of Petroleum, and Sonangol. Their individual roles can be briefly described as follows:

- *Ministry of Finance (MOF).* The DNI (National Tax Inspectorate) in the MOF is responsible for the assessment and collection of royalties and taxes. The office is very small (three to four professionals) and while it does a good job given its resources, it must rely heavily on outside auditors in performance of its assessment function. The DNI engages qualified international auditors through international competitive tender every three to four years. Current auditors are Deloitte Touche. Estimated taxes are paid monthly, 30 days after the end of the month in question. Once the annual audit is complete, a final adjustment is made for any over- or under-payment in March following the end of the tax year. DNI is responsible for recording all tax transactions and regularly publishes tax assessments in considerable detail on its web site (www.minfin.gv.ao). Details of the petroleum tax cycle and procedures are shown in Box 3.3.
- *Ministry of Petroleum (MOP).* The MOP is responsible for determining the prices to be used for fiscal purposes. Procedures are set out in each concession decree. Contractors provide price information on their own transactions to the MOP. On the basis of these submissions, the MOP determines a market price (MOP) for the period (quarter) in question. In the event of disagreement, the Concession Decree provides for recourse to an independent expert opinion, which the minister must take into account in making a final decision. The system is generally regarded as working well, resulting in prices reflective of arm's length market prices.
- *Sonangol.* As concessionaire, Sonangol oversees taxpayer cost recovery claims, engaging qualified external auditors to do this, much as the DNI contracts out its tax assessment functions. At the moment, cost audits are conducted by Ernst and Young (others are engaged where conflict of interest situations arise). As concessionaire, Sonangol is also responsible for determining profit oil due to the government, for its marketing on behalf of the State and for payment of the proceeds thereof to the MOF (see Table 3.3).

> **Box 3.3: The Petroleum Tax Cycle in Angola**
>
> Angolan tax legislation employs the term liquidation of tax, which essentially means quantification of the amount of tax payable. Liquidation of oil taxes other than Training Contribution must be made at the competent tax office (Article 59.1).
>
> *Main Oil Taxes—Provisional Liquidation and Payment.* Under Article 59.2 taxpayers are responsible for making a provisional liquidation, based on their provisional tax declaration, of PPT, PIT, PTT and National Concessionary earnings (which should be taken as including the price cap excess payable under the early PSAs). With two exceptions, the provisional liquidation must take place in the month following the month oil is lifted. The exceptions are that PIT and PTT payable under the equity partnership regime must be liquidated in 12 equal monthly installments in the course of the fiscal year. The provisional tax declaration must be reviewed every quarter and amended as necessary, so in practice the installments will be increased or reduced to take account of these revisions. Payment of these taxes must be made within the same time limit as the provisional liquidation (Article 62.1).
>
> *Main Oil Taxes—Final Liquidation and Payment.* Under Article 59.3 the final liquidation of these oil taxes must be made in the same month as the final tax declaration, i.e., the month of March following the end of the fiscal year. Payment of the final tax due must then be made within 30 days (Article 62.2). If the final declaration shows that tax has been overpaid, companies should deduct the overpayment from their next payment(s) of tax.
>
> *Payment of Surface Tax.* The date for payment of Surface Tax is set out in Article 53. For concessions within the equity partnership regime, it is payable annually in the month following the date when the concession was granted (e.g., if the concession was originally granted on November 14, 1980, Surface Tax would be payable each year within 30 days of November 14.) For concessions within the PSA regime, Surface Tax for each development areas is payable annually within the month following the date of the original commercial discovery.
>
> *Payment of Training Contribution.* Payment of Training Contribution is regulated by separate government decree. Article 57.2.
>
> *Payment of Additional Tax.* Where further tax becomes payable (e.g., following determination by the Fixation or Revision Commission), payment must be made within 15 days of the notice of liquidation (Article 62.3).

Governance, Transparency and Institutional Capacity

Appropriate management of oil wealth in Angola will require improvements in the areas of governance, transparency and institutional capacity. As affirmed elsewhere, in an oil economy, the presence of weak institutions may invite rent seeking and waste. In the case

Table 3.3. Sonangol Tax and Profit Oil Liabilities to the Government of Angola

Year	($US million)
2000:	1190
2001:	721
2002:	642
2003:	821
2004:	994

Source: Sonangol's own assessment as reported by the IMF. Includes upstream activities only.

> **Box 3.4: The Paradox of Plenty and the Case of Angola**
>
> Four principal contributors to the Paradox of Plenty have been identified. Two of these are technical—the so-called "Dutch Disease", and oil revenue volatility. Two are more political in nature—weak governance and the lack of and/or failure to develop the institutional capacity required to the challenges of resource wealth.
>
> **Dutch Disease.** This "disease" is named for the problems experienced by the Netherlands following the discovery and initial exploitation of vast domestic reserves of natural gas. Large scale revenue inflows from oil exports put upward pressure on the exchange rate. They also lead to a significant expansion in domestic demand relative to the country's ability to supply that demand. The demand expansion comes from the budget and public sector and, where oil revenues get into the domestic banking sector, from credit expansion. The demand expansion in turn increases the price of non-traded goods, causing a further appreciation of the real exchange rate. The combination of these two impacts results in an often dramatic decline in the competitiveness of non-oil exports, a shift in domestic resources away from those sectors to the non-traded goods sectors, and erosion of diversity and balance in the domestic economy. Evidence of the Dutch Disease has been identified in almost all countries where petroleum exports play a major economic role, including in Angola.
>
> **Oil Revenue Volatility.** The oil industry is notorious for its often violently cyclical behavior. This may be due to the uncertain pace of oil discovery and development. At a global level, however, the cyclical character of the industry is more often attributable to the volatility of oil prices. Volatility makes economic management difficult in itself, especially as cyclical swings are typically not predictable. Difficulties are compounded by the positive correlation commonly observed between revenues received and expenditures. Volatility in revenues, associated with volatility in aggregate expenditure, public and private, creates real exchange rate volatility. This volatility will make profits in the tradeable sector very unstable and risky, creating further disincentives towards investment in these sectors.
>
> **Governance.** Good governance is widely recognized as critical to successfully addressing the Paradox of Plenty. Good governance is variously defined, but it is clearly multi-dimensional and should include, among other things: clear and stable laws and regulations; rule of law; a high level of competence in government; fiscal, budgetary and monetary discipline; public-private sector balance in the economy; an open dialogue between government and civil society; and a high degree of transparency. Unfortunately, oil-rich developing countries do not score well against these governance indicators (see analysis in Chapter 5). Weak governance in many of these countries probably pre-dated oil development and made it difficult to manage oil wealth from the outset. However, a range of arguments and evidence exists which suggests that the arrival of significant oil wealth can itself undermine governance creating a vicious cycle.
>
> **Institutional Capacity.** Oil revenues often exceed an oil-rich country's capacity to manage them effectively, or to ensure their efficient investment. The result is often macroeconomic mismanagement and waste on a major scale. In addition to straining, or even overwhelming, existing institutional capacity, oil wealth may actually erode incentives to invest in creating an effective civil service. Government decision makers may feel little need for a skilled administrative team to handle oil revenues when so much money appears to flow so easily from a very concentrated source. More insidiously, those benefiting most from oil may perceive an effective, efficient and watchful civil service as a threat to the benefits they enjoy. As suggested above, adequate institutional capacity is really a component of good governance. It is commonly singled out for discussion, however, because of its special importance. This is certainly the case for Angola.

Source: World Bank (2005), *Managing Angola's Oil Wealth,* Background Paper for the Country Economic Memorandum, Washington, D.C.

of Angola, where the level of institutional quality is low, in part because of the effects of the war and in part because of problems associated with the Paradox of Plenty (as discussed in Box 3.4), it is extremely necessary to improve the factors that can affect the quality of governance. This will require further efforts to increase transparency in the sector and develop institutional capacity in the Government to manage the country's oil wealth appropriately. This section of the report discusses the role of Sonangol in the economy and its relationship with the Ministry of Finance, recent developments in the areas of governance and transparency, and highlights areas where faster progress needs to be made.

The Role of Sonangol

Sonangol performs multiple roles vis à vis the Government, including activities which would normally be performed by the Treasury and the Central Bank. It is a taxpayer, it carries out quasi-fiscal activities, it invests public funds, and, as concessionaire, it is a sector regulator. The quasi-fiscal operations assigned to the national oil company by Government include: a) supply of petroleum products to the internal market at subsidized prices; b) product deliveries below cost to public services (military, hospitals, etc.) and in remote areas; c) management and servicing of public debt; d) the performance of various sector regulator roles; and e) marketing of the Government's share of crude oil accruing to Government under PSAs. This gives the company a largely independent role in the economy which could be justified in a war period, but not in the context of peace.

This complicated relationship between the Government and Sonangol has resulted in payment arrears to the Government and in the appearance of a dual spending system. At the same time that Sonangol performs all of the operations described above, it is also subject to the taxes and other payment obligations, including production shares, as any other oil company in Angola. Sonangol's record, however, is one of underpayment of taxes and substantial arrears to the Government. This is because the company is often not reimbursed for the cost of carrying out these tasks and, in lieu of reimbursement, the company has offset or claimed credit against its tax liabilities by the amount of such costs. The practice has resulted in a dual spending system associated with the spending executed by Sonangol and by the National Treasury.[33] In an ideal situation, Sonangol should have to transfer to Government what it owes to Government, and the Government would be the one responsible for undertaking many of the quasi-fiscal expenditures which are currently performed by Sonangol.

Independently of how organized these non-conventional mechanisms have become, they violate the existing financial legislation. While during 2002 the compensation process described above seemed to be rather haphazard, since 2003 it has become more organized and predictable. This in itself presents the additional danger of creating an artificial "functionality" that could lead to the perpetuation of a mechanism that violates basic legislation and ultimately weakens the Ministry of Finance as the chief fiscal institution in Angola. In addition, the existence of a significant amount of expenditures which are not timely recorded according to the current procedures followed by other spending units in the government weakens the budgetary process and creates uncertainty as regards the actual fiscal stance of the Government. It also weakens transparency and accountability, and impairs planning.

---

33. A detailed assessment of this offsetting mechanism is described in the World Bank's PEMFAR report published in February 2005.

More needs to be done to phase out the dual spending system. Although there have been initiatives to add transparency in the relations between Sonangol and the Government, the current procedures have several disadvantages, in addition to that of recording fiscal activities outside the budget where they belong: a) as affirmed above, it is not transparent; b) it leads to frequent disputes; and c) it misrepresents Sonangol's position on taxes. The GoA is now moving to reflect quasi-fiscal activities in the budget, and Sonangol is taking steps in its external audit to identify and audit all such activities undertaken on behalf of the Government with a view to resolving tax disputes as well as clarifying the financial results for its core activities. Both of these measures are welcome. However, the offsetting mechanism is still very much a current practice and concrete efforts to phase it out have yet to materialize.

Sonangol can contribute more effectively to the development of the country by focusing on its own business. It is a recognized fact that the activities performed by Sonangol on behalf of the government create an additional administrative and operational burden on the company. If these were relinquished and Sonangol focused exclusively on its own business, the company could contribute even more to the development of the country. A few examples of areas where the increased intervention of Sonangol could make a difference include: (a) investments to increase fuel storage capabilities; (b) investments to improve fuel distribution to the interior of the country; (c) training of Angolan labor force to work in the IOCs; and (d) investments in social projects.

Governance and Transparency

There has been some progress in improving governance and transparency in the petroleum sector since 2002. A number of recommendations made by the Bank and the Fund, both in the Oil Diagnostic Study and in the PEMFAR, have been adopted and these are summarized below.

- The government has been regularly publishing details of oil payments received (by block, by type of payment, with annual summaries by company) on the Ministry of Finance website, with a lag of 6 months. This level of published detail is virtually unique among oil producing countries.
- The audits of the petroleum sector have been conducted on a regular basis. The financial statements of companies in the Sonangol group were audited comprehensively for the first time, by Ernst & Young, in 2003, with only minor audit qualifications of exploration and production operations (where most of the money is). The less important downstream operations did, however, attract numerous audit qualifications, but Sonangol has invited Accenture to help clean up irregularities. The auditors' reports, however, have yet to be published.
- Recent audits have followed acceptable accounting rules. The audits mentioned above applied Angola's recently issued accounting legislation where appropriate, and IAS rules where the new legislation does not apply. The new audit legislation is already regarded as a major step towards IAS accounting. By end-2006 Sonangol expects to have moved fully to IAS standards. The 2004 audit was recently completed.
- The roll out of the IFMS (SIGFE) is progressing steadily. Progress towards the implementation and full roll out of the Angolan IFMS (Integrated Financial

Management System) has been steady and currently most of the expenditure side of the government accounts is registered in the system. There are delays associated with the inclusion of all revenues in the system, but the Ministry of Finance has a clear plan to address the constraints and the full roll out of the system is expected to take place by 2008.
- A model is being used for oil revenue forecasting. The Ministry of Finance has a signed a contract with AUPEC, Aberdeen University Petroleum Economic Consultancy, to implement an oil revenue forecasting model and advise DNI on strengthening its capacity with respect to petroleum taxation. The work started on April 1, 2006 and Ministry staff are being trained on how to use the model.

Other structural financial management issues are being addressed, but progress has been slow. The Government has started to strengthen the capacity of the Ministry of Finance to control expenditures through the roll out of the SIGFE and ring-fencing the operations of Sonangol on behalf of the Treasury. But more needs to be done to arrive at a point of "normalization". Other suggested improvements in public financial management that remain to be addressed include:

- Sonangol's quasi-fiscal operations: Sonangol's quasi-fiscal operations are currently reflected in the budget, but there is no clear indication about a timeframe for phasing them out. The Ministry of Finance has confirmed that the expenditures performed by Sonangol on behalf of the government comply with the formal budgetary procedures, but that this is currently happening with a lag of 90 days. The National Directorate of Accounting (DNC) has confirmed to the staffs of the Bank and the Fund that their goal is to eliminate this lag completely by the end of 2006.
- Separation of concessionaire and operator roles of Sonangol. The Government and Sonangol have both indicated that there will be no change on this configuration at least until 2010 under the justification that there are institutional and technical limitations in the Ministry of Finance and in the Ministry of Petroleum that prevent faster progress. The Oil Revenue Diagnostic Study jointly funded by the Bank and the Government recommended transfer of the regulatory functions away from Sonangol.
- Updating of the Ministry of Finance website. Detailed information was provided to the Bank and Fund in March 2006 about oil exports, oil prices, and profit oil from the Tax Directorate of the Ministry of Finance for 2005, but the information on the website continues outdated (reflecting the situation in the first quarter of 2005 only when the accounts for 2005 have already been closed).
- Engagement with civil society on public finances management issues. The MOF has asked the World Bank to organize two high level workshops on petroleum revenue management—one for senior Government officials, followed by a second for industry and civil society. These workshops were successfully held in mid-May, 2006. Other outreach activities for civil society should be organized in the future.
- Formal endorsement of the EITI criteria. Angola has been cautious about announcing formal adherence to the principles and objectives of the Extractive Industries Transparency Initiative (EITI), despite the encouragement of the Bank,

IMF and bilaterals. Nonetheless, the government has made some progress towards addressing the criteria of the EITI and the authorities should seriously consider joining the initiative.

A scorecard of actions taken and targets to be met spotlights required further actions. Table 3.4 describes areas where progress has been made and where further actions are

Table 3.4. A Scorecard to Assess Governance and Transparency in the Oil Sector

Criteria	What Has Been Done So Far	Required Further Action
Resolve conflict of interest potential (Sonangol as concessionaire)	Sonangol is ring-fencing concessionaire activities	Subject to credible institutional capacity, transfer concessionaire role to Ministry of Petroleum
Adequate Government oversight of Sonangol	MOF/MOP have little capacity to provide oversight	Engage qualified consultant support. Build capacity.
Reconcile/include Sonangol financial flows with budget	Sonangol ring-fencing and auditing quasi-fiscal activities. Own expenditures are still outside budget. Quasi-fiscal expenditures comply with budget procedures with a 90-day lag/	Bring Sonangol quasi-fiscal and own expenditures into budget and comply with budget procedures without delays.
Qualified, independent audit of payments made and revenues received	Annual industry cost and fiscal audits by experienced international auditors	Take notice and act according with the auditors' recommendations. This will allow a comparison of payments made by industry and revenues received by the federal and provincial governments. Audit revenues received by MOF, Cabinda and Zaire provinces.
Publication of audit results in accessible form	Current detailed publication of company payments on MOF website. Audit results not yet published.	Add audit results. Improve accessibility of website. Consider broader media publication.
Audit exercise applies to all companies, including Sonangol	Current practice, but Sonangol data derived from block operator.	Sonangol to provide data directly to MOF. Publication of Sonangol corporate audits
Engage civil society in revenue management and transparency process	No current engagement	Topical workshops to include civil society. Establish independent public information center
Clarity on legal, contractual and fiscal framework/ procedures	Legal drafts and texts difficult to access. DNI preparing tax manual	Compile and publish legal texts, procedures. tax manual.
Develop time-bound, funded, action plan for implementation of transparency agenda	No current plan, although individual components have been scheduled	Prepare and publish explicit plan.

required to meet objective governance and transparency criteria which are considered good practice.

Despite recent progress, there are critical outstanding items that remain to be addressed. The Government acknowledges that there are limitations and constraints to moving on a few areas, and has asked the Bank to provide advice on how to overcome them. While discussions on a definitive list of performance criteria are ongoing, a number of key areas where the government will need to be proactive are summarized below:

- Formal endorsement of EITI and agreement on an action plan for its implementation.
- Clearing of tax arrears by Sonangol.
- Verification of all Sonangol liabilities and payments.
- Publication of audits of financial statements and tax audits.
- Extension of the audit of company results to audit of the "flip side", i.e., of revenues received by federal and provincial governments.
- More accessible, user-friendly website for oil data, with additional data/informational enhancements (e.g., tax-payers manual).
- Publication of the Sonangol auditors' reports.
- Outreach informational events for civil society.
- Public Information Centre with objective information on the oil industry.

Oil Wealth: How Much and for How Long?

Besides addressing governance and transparency issues, it is equally important to assess how much oil wealth will be accumulated and how long it is likely to last. Gaining a good understanding of the size of the oil wealth that the country will accumulate in the years to come is essential to plan for the future and to define a strategy to improve the lives of the Angolan people. The report now turns the focus to these issues and to policy options available to the authorities on how to manage their rapidly growing oil revenues.

In the computation of Angola's oil wealth, three reference price scenarios are considered (a base case, a high case, and a low case). Benchmark price scenarios were developed by the World Bank's Commodities Department and are illustrated in Figure 3.5. The benchmark used is the widely referenced price of Brent North Sea oil, which applies to Angola. It is also worth noting that the price of Angolan oil varies with the quality of the oil extracted in the different blocks (see Box 3.5). The rationale for each of one of the different scenarios is as follows:

- In the base case scenario, the thinking is that oil prices are expected to remain elevated in the near- to medium-term due to capacity constraints, moderately strong demand growth, and OPEC production discipline. High oil prices are expected to have some impact on demand, and non-OPEC supplies outside of the Former Soviet Union, which have been flat the last few years, are expected to rise from such areas as the Caspian, West Africa, deepwater, and Canada. Output in Russia is also expected to rebound. New capacity is expected from most OPEC countries such that surplus capacity will rise modestly and exert downward pressure on prices. However, OPEC will be in a fairly strong position to set a reasonable floor under prices.

Figure 3.5. Brent North Sea Oil Price Scenario

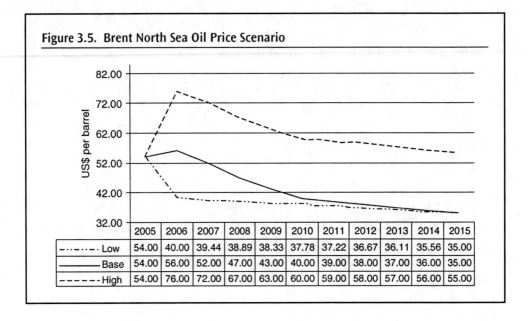

	2005	2006	2007	2008	2009	2010	2011	2012	2013	2014	2015
Low	54.00	40.00	39.44	38.89	38.33	37.78	37.22	36.67	36.11	35.56	35.00
Base	54.00	56.00	52.00	47.00	43.00	40.00	39.00	38.00	37.00	36.00	35.00
High	54.00	76.00	72.00	67.00	63.00	60.00	59.00	58.00	57.00	56.00	55.00

- The high case scenario assumes stronger growth in demand, particularly in China and India, and continued supply disappointments that keep the market balance tight. High prices ultimately have a negative impact on demand and a positive impact on supply such that prices fall over the forecast period. However, OPEC is in a strong position to keep prices above $50 per barrel.
- The low case scenario assumes that demand weakens from already high prices and structural changes in demand going forward. Supplies grow rapidly in both OPEC

Box 3.5: The Quality of Angolan Crude Oil

Angolan crude oil is generally of high quality, medium to light gravity (30° to 40° API) with low sulfur content. The most commonly traded crude oils are Cabinda, a crude oil blend, Soyo, and Palanca blends, and the Nemba, Kuito, and Girassol crude oils. In some few cases, Angolan crude oils sell at a discount off Brent prices due to quality disadvantages, and, more generally, to locational disadvantages. Discounts applied to the Brent price for the purposes of analysis in this paper are taken to be as follows:

Soyo	−1%	*Kuito*	−15%	*Canuku*	−1%	NGL	−4%
Palanca	−1%	*Landana*	−5%	*Xikomba*	−5%	Soyo	−1%
Kiame	−7%	*Hungo*	−7%	*Negage*	−5%	Nemba	Parity
Girassol	Parity	*Benguela*	−7%	*Block 31*	−5%	LPG	−15%
Dalia	−9%	*Plutonio*	−5%	*Cabinda*	−4%		

Source: Wood-Mackenzie's GEM.

Table 3.5. Angola's Petroleum Wealth under Different Price Scenarios

Asset	Low price US$ Mi	Base price US$ Mi	High price US$ Mi
Current proven and probable assets	118,597	145,718	253,310
Current assets adding Block 31SE and 32	123,378	150,657	266,027

Note: Assets discounted at 10 percent and reported in $ million.

and non-OPEC countries, exerting significant downside pressure on prices. OPEC takes a less aggressive stance with respect to prices to maintain market share, but is able to secure a price much higher than in the past at around $30 per barrel in nominal terms. Angolan crude oils sell at a discount off Brent prices due to locational disadvantages and, in some cases, quality disadvantages.

Angola's future oil revenues are substantial and estimates show that they can reach close to US$266 billion in present value terms. The present value of that asset, discounting future revenue flows at a 10 percent rate, results in estimates ranging from $119 billion to $266 billion, depending on the production and price scenarios assumed (see Table 3.5). These figures translate into per capita petroleum wealth ranging from $8,500 to $19,000, based on an estimated population of 14 million. While large, these figures are perhaps lower than many policymakers might have expected. This is because of the discounting and the relatively short-lived peak of production, even in the extended production case. In what follows, the report looks into the future path of oil production and its impacts on government revenues under different scenarios.[34] The reserve, production, cost, and fiscal scenarios used to project oil revenues are those described in preceding sections. The recurring critical component is price.

Oil Revenues under Different Scenarios

Gross oil revenues are expected to increase quickly over the next five years and then should decline sharply thereafter in the absence of new discoveries. Gross oil revenues have grown rapidly over the past several years reflecting both price and production escalation. A steeper increase in gross revenues is expected in the near term. Figure 3.6 and Table 3.6 summarize past and forecast gross revenues under the three assumed price scenarios. The sums involved in each of the three scenarios are large. A sharp peaking in revenues, led by PSA production, is expected over the next five to ten years, with an equally sharp fall thereafter, reflecting expected price moderation and production decreases. The fall may be deferred by bringing the pending development projects in Block 31 and 32 onto production, but

---

34. Our projections do take into account new discoveries. In fact, a new bidding round for exploration of existing fields has been announced, but any significant new discovery, should one be made, could not be expected to impact revenues positively for at least 5 to 7 years.

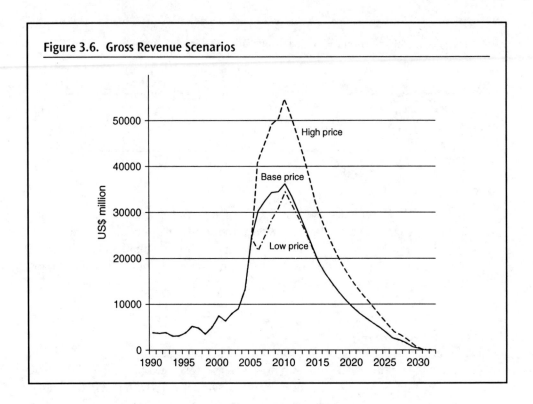

Figure 3.6. Gross Revenue Scenarios

this would extend the period of high revenues by no more than two to three years. While all revenues are significant, a comparison of the three scenarios illustrates their dramatic volatility in the face of not unrealistic future price scenarios.

The share of the government in total oil revenues is also expected to grow quickly and then decline steeply in the near future in the absence of new discoveries. The past and expected future behavior of government revenues is illustrated in Figure 3.7 and Table 3.7. Government revenue closely tracks gross revenues with a slight lag reflecting early recovery

Table 3.6. Gross Revenue Scenarios

Price Scenario	Gross Revenues (US$ Million)		
	Base	High	Low
1990–1994	17,340	17,340	17,340
1995–1999	22,026	22,026	22,026
2000–2004	43,884	43,884	43,884
2005–2009	156,052	210,063	129,852
2010–2014	148,768	226,698	143,266
2015–2019	73,388	115,324	73,388
2020–2024	35,152	55,239	35,152
2025–2029	11,018	17,314	11,018

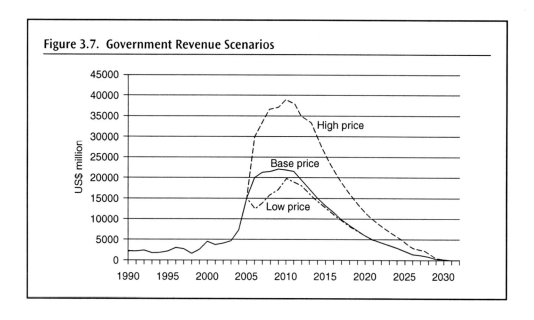

Figure 3.7. Government Revenue Scenarios

of costs by contractors and a consequent deferral of payments to government. Also, as might be expected, based on the fiscal regimes described earlier, government revenues increase as a percentage of project net cash flow (gross revenues minus cost) in later years following cost recovery by the contractors. They also increase as prices increase, e.g., they are higher as a percentage of project cash flow in the high price case than in the low price case.

Government revenues will also vary in line with Sonangol's revenues and expenditures. The measures of government revenue shown above do not include the after-tax revenues accruing to Sonangol. As Sonangol is 100 percent state-owned, these are public revenues and should be considered together with other revenues in describing total revenue flows to government from the sector. Sonangol revenues expected under each of the three

Table 3.7. Total Government Revenues

	Total Government Revenues (US$ million)		
	Past revenues at historical prices		
1990–1994	10,247		
1995–1999	12,119		
2000–2004	24,251		
	Future revenues under three price scenarios		
	Base price	High price	Low price
2005–2009	99,930	152,358	74,301
2010–2014	95,016	175,672	86,805
2015–2019	50,239	94,850	48,232
2020–2024	21,556	41,922	21,256
2025–2029	5,772	11,748	5,714

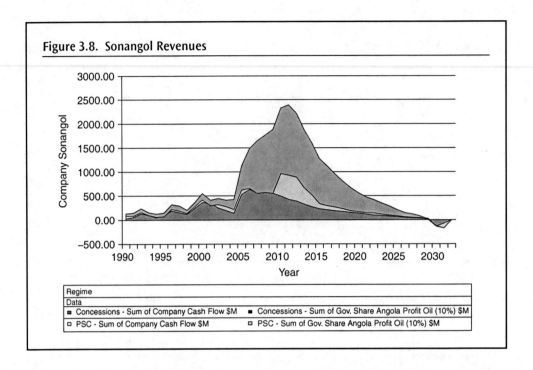

Figure 3.8. Sonangol Revenues

assumed price scenarios are described in Figure 3.8. The revenues comprise equity returns from both concession areas and PSAs in the form of after-tax cash flow and contractor production shares, respectively, and the 10 percent of government production shares allocated to Sonangol to cover costs incurred in discharging its duties as concessionaire. Sonangol not only receives significant public revenues, but it also spends public funds, reducing government's net revenues from the oil sector. Sonangol's expenditures are expected to grow in the near term as its commitments to fund its annual share of deepwater development projects become effective, as shown in Figure 3.9.

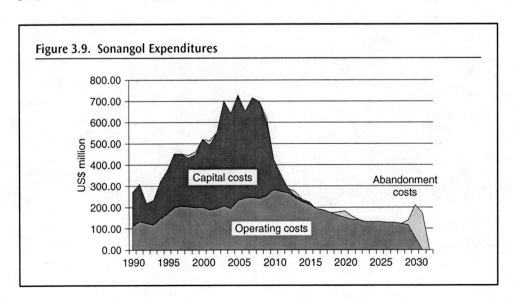

Figure 3.9. Sonangol Expenditures

Intergenerational Considerations

With the massive increase in fiscal revenues associated with the upcoming oil boom lies the question of how to allocate the proceeds of the oil windfall in an intertemporal perspective. That is, what is the optimal annual level of consumption out of Angolan oil income now and in the future? The answer depends on both the size of the revenue that can be expected from oil in the future, the rate at which one discounts these values to the present, and the relative weights one attaches to future versus present generations. Past attempts in oil-rich countries to address these intergenerational concerns have almost uniformly run into difficulties. Problems have occurred at the political level—it is hard for politicians to "sell" the idea of transferring resources to future generations when the needs of the current generation are manifestly great. Additional problems have had their origin in the weak governance and institutional capacity traceable to the Paradox of Plenty.

To guarantee the well-being of future generations, part of Angola's oil revenues should be saved. Based on the current reserves and development activity described in previous sections, the current oil boom in Angola is likely to last for about two decades *in the absence of new discoveries*. During this period the economy will experience high economic growth, driven by the oil sector, and a massive windfall with concomitant fiscal gains. Provided the serious issues associated with the Paradox of Plenty (see Box 3.4) are adequately addressed, the fiscal gain should go a considerable way towards alleviating Angola's widespread poverty, contribute to growth and economic diversification, and increase the general standard of living. However, given the expected decline in production from peak boom levels and the eventual exhaustion of oil resources, long term sustainability of these benefits requires that a part of the coming revenue windfall be saved to provide for the economic and social well-being of future generations.

An option to make the oil wealth last longer is to convert part of it into financial assets that can yield returns well beyond the depletion of oil reserves. In order to transform the revenue boom into a permanent income per capita, Angola can convert oil revenues into financial assets at a rate that allows the return on those financial assets to take over as a source of income in the long run. In such a scenario, Angola would invest, as oil revenues come in, the amount required for interest income to equal oil revenues per capita at the end of the oil boom. For sustainable expenditure policy, this representation of oil wealth is more realistic than the absolute numbers, as reported in Figure 3.10.

Future government revenues and expenditures under this approach would vary with the rate at which financial assets are remunerated. Based on projected future revenue flows, the permanent sustainable expenditure per capita is $169 per year. This means that each Angolan citizen could spend an additional $169 per year in perpetuity if Angola saves all oil revenues that exceed this amount for as long as oil revenues are coming in. In calculating this number, we assume an annual population growth of 2.9 percent, the base price scenario in Figure 3.10, an annual yield of 5 percent on financial assets, and we discount future revenue flows by 10 percent per year in order to include uncertainty into the calculation. Different assumptions give different results: if we discount future revenues by 15 percent and assume a low price scenario, the sustainable permanent expenditure is $107 per capita. If we discount revenues by 5 percent and assume a high price scenario, the best case sustainable permanent expenditure is $385 per capita (see Table 3.8).

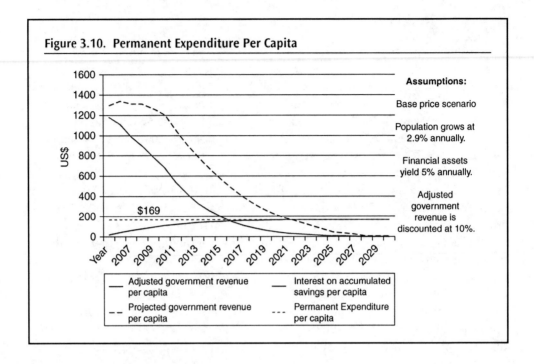

Figure 3.10. Permanent Expenditure Per Capita

Policy Options To Manage the Windfall

Fiscal policies will be of central importance in managing looming revenue shocks. As revenues are expected to increase significantly under all likely scenarios, policy objectives should include dealing with Dutch Disease risks (avoiding overheating of the economy in response to massive revenue inflows and consequent price inflation, dealing with the unavoidable real exchange rate appreciation and likely erosion of the competitiveness of non-oil export activities), avoiding damaging pro-cyclical expenditures in response to expected revenue swings, and providing for expected declines in production and revenues, and ultimately exhaustion of Angola's oil resources. In Angola, the desirable fiscal policy should attempt to insulate the economy from the volatility of oil revenues, because frequent upward or downward adjustment of fiscal expenditures are costly and hurt the economy through uncertainty about aggregate demand and through costs associated with factor reallocations.[35]

Econometric estimates for the case of Angola show that controlling oil revenues seems to be essential to effectively control the growth of government spending and insulate the economy from oil price volatility. A detailed econometric analysis on the relationship between oil revenues, government expenditures, and gross domestic output in Angola shows that following an increase in oil taxes, government spending temporarily increases and then fades away quickly after the shock (Carneiro 2005). In the same token, GDP responds positively to an increase in oil revenues but this effect is not sustainable in time. In this context, in which government spending and GDP vary in the same direction as oil revenues, the introduction of some sort of stabilization fund to insulate the economy from

---

35. See Katz and others (2004) for a generalization of this point.

Table 3.8. Permanent Expenditure Per Capita at Different Assumptions

Price Scenarios	Discount Rate		
	5%	10%	15%
Low	182	136	107
Base	220	169	136
High	385	292	232

the volatility associated with the oil sector could be warranted.[36] As argued by Barnett and Ossowski (2003), a key policy objective in such situation should be to pursue a fiscal strategy aimed at breaking the pro-cyclicality response of the fiscal policy to volatile oil prices. This would involve eliminating expansionary fiscal policy biases in times of oil booms.

Up to now, the enclave nature of the Angolan economy has meant that GDP has responded more to changes in oil revenues than to changes in government spending. The econometric estimates also show that oil revenues and GDP do not respond significantly to increases in government spending in the short run.[37] One possible reason why short-term changes in government spending do not affect GDP could be that government spending simply leaks out through the trade balance due to the non-existence of idle capacity which makes domestic supply inelastic. That is, if all the government is doing with its additional expenditures is financing imports then one would not see GDP changing. The finding is not just consistent with the enclave nature of the oil sector in the Angolan economy and the prevailing Dutch-disease situation that plagues Angola and other resource-rich countries,[38] but could also be seen as a consequence of the prolonged civil war that plagued the country for over 27 years (see Collier 1999).[39]

The Way Forward: Five Steps to Manage the Windfall

There are five important issues that deserve attention. A crucial question behind the extraction of natural resources that are exhaustible is how to manage a sudden increase in government revenues derived from windfall gains. In the case of Angola, there are five associated issues that call for careful consideration by the authorities.

First, the authorities should strive to improve the investment climate and the business environment in Angola to foster the gradual and sustainable growth and diversification of the non-oil economy. The extraction of the country's oil reserves by definition represents

---

36. There are certain pre-conditions that must be observed for a successful implementation of a petroleum revenue fund and an assessment as to whether these are present or not in Angola is beyond the scope of this paper. For a thorough discussion of this subject, please refer to Davis and others (2003) and Katz and others (2004).
37. A result which has also been found for the case of Mexico (Tijerina-Guajardo and Pagán 2003).
38. For an assessment of different ways in which the Dutch disease can operate in the economy, see Hausmann and Rigobon (2003).
39. It is reasonable to assume that if government demand is channeled to the domestic economy by means of increased investments in infrastructure and human resources, then the non-oil domestic sector will in the medium to the long run eventually respond, leading to increases in GDP.

a depletion of its natural resource endowment that will affect not only the current generation. Clearly, the country receives something in return, but from a long-term sustainability point of view, this reduction should ideally be compensated by the accumulation of other asset forms such as physical and human capital. The decision on the fraction of oil revenues to be saved as financial assets, therefore, requires that policymakers make explicit decisions about the intergenerational distribution of revenues related to the extraction of exhaustible oil resources. In particular, future generations will be worse off if the oil-related revenues are spent too quickly, without leading to improved long term non-oil growth prospects. Thus, the authorities should strive to create the conducive business environment in Angola that will support sustainable growth and the diversification of the non-oil economy. Options available on that front are discussed in detail in Chapter 5.

Second, the fiscal envelope for the coming years should be based on a medium-term economic framework. Currently, the horizon for decisions that affect macroeconomic policy is, at mot, two years ahead as spending programs in annual budgets are determined only by revenue prospects for the coming year. Despite the fact that the government has been adopting a conservative approach towards specifying the oil price on which revenue projections are based, this approach can still yield sharp cycles in spending which can prove unsustainable in the long run. A preferred approach would be to base annual ceilings for public expenditure on medium-term revenue prospects, evaluated at a long-run oil price. Under this approach, consistent fiscal surpluses could be generated over time so that financial reserves could be accumulated during the years of peak oil production to sustain incomes when oil revenues eventually fall.

Third, fiscal policy should be anti-cyclical in relation to oil prices. The pro-cyclicality of fiscal expenditures and oil prices is dangerous and can transmit volatility to the rest of the economy. In countries that have not been cautious, the resulting fiscal deficits have been financed with external and/or domestic borrowing. The former type of borrowing has rendered many borrowers vulnerable to increases in the interest rate on foreign loans, as well as to the drying up of new loans as sustainability concerns set in, while the latter has often been inflationary or has crowded out private sector access to credit. The joint result could well mean that the government will be forced to adopt belated, costly, and disorderly expenditure cuts in the future, often involving the suspension or abandonment of investment programs. The bottom line for these governments is that in periods of oil price downturns the authorities would be forced to adopt sharp and disruptive fiscal contraction measures when the economy can least afford them.[40] Therefore, a key policy objective in this context should be to pursue fiscal strategies aimed at breaking the pro-cyclical response of expenditures to volatile oil prices. In this sense, the approach of using a conservative oil price in the preparation of the budget that the authorities are already adopting is welcome, but this approach should be complemented with the recommendation of the previous paragraph.

Fourth, the authorities should build up foreign exchange reserves and reduce the country's public external debt. In view of the improvements in Angola's fiscal position,

---

40. For a discussion of procyclical fiscal policies in Nigeria and Venezuela, for example, see Hausmann, Powell, and Rigobon (1993), and García et al. (1997).

there is considerable scope to reduce the country's public external debt while also building up international reserves. Part of the windfall gains that is accruing to the government can be used to clear external debt arrears, which currently add up to close to US$2 billion dollars plus late interest, and an important first step in that direction could be the resumption of payments falling due. The authorities should also consider rebalancing the overall foreign exchange portfolio with a view to maximize returns to Angola and eliminate the government's reliance on expensive oil-backed loans from commercial banks and to minimize recourse to external debt to finance public investment in the future.

Finally, expenditures should not rise faster than transparent and careful procurement practices allow. Massive institutional and legal reforms are necessary to develop a thriving business environment and in Angola these reforms have started but are in a very early stage. In simple terms, the level of spending should be determined taking into account its likely quality and the capacity of the administration to execute it efficiently. In this regard, an abrupt enlargement of expenditure programs associated with oil windfalls carries important risks. A hasty public spending program may exceed the government's planning, implementation, and management capacity, with the result that it may be difficult to prevent wasteful spending. For instance, the criteria for the selection of capital projects may become lax, leading to suboptimal choices; the costs of new projects may also increase due to bottlenecks in the supply of some inputs; and a large-scale capital expenditure program can also be a fertile ground for governance problems.

Institutional Options to Manage the Windfall

An institutional framework will be required in order to effectively manage Angola's oil wealth. Regardless of whether or not Angola adopts this guide to fiscal policy, the Government should have the institutional capacity required to generate future revenue scenarios. Considerable progress has already been made in this direction in the form of the Oil Revenue Diagnostic Model prepared by AUPEC for the MOF, but additional arrangements are still required such as, for example, in opening up the dialogue with civil society and in establishing a formal and transparent mechanism to administer the windfall.

Popular support for saving a share of oil wealth can be gained through an open dialogue with civil society. The permanent expenditure approach to fiscal policy discussed above is likely to encounter significant popular/political opposition because of the substantial early savings/deferred expenditures it implies in years when the perceived need for public expenditures is high. To gain popular support, however, the authorities need to strengthen the dialogue on oil revenue management options with civil society and make the case that the permanent expenditure approach establishes a benchmark that investments must meet in order to convert the oil boom into sustainable wealth.

The establishment of the savings and expenditure rules for the oil reserve account should draw on the success and failure of past experience in other oil producing countries. The authorities have announced that part of the improvement in the country's fiscal position will be used to create a dedicated oil reserve account of the Ministry of Finance at the BNA to help in fiscal stabilization. Past experiences with similar initiatives in a variety of countries have not been encouraging. However, learning from past failures, new promising approaches are being adopted around the world. As the Government of Angola is considering the establishment of an oil reserve account, the following features, which are based on other models now being developed and implemented in other countries, might be considered useful on the way forward:

- The oil reserve account should be the designated recipient of petroleum windfall revenues. Partial capture of oil revenues by various ministries or state-owned entities would make it difficult to obtain efficient and sustainable management of the revenues.
- Transfers to the budget should be based on the agreed savings/expenditure rule. Without clear rules, transfers risk becoming increasingly dependent on short-term economic and political cycles, which would critically undermine the account's function.
- The BNA should be designated as the operational (day-to-day) manager for the account. An efficient, transparent, and rule-bound management requires dedicated and neutral staff, placed within the Central Bank.
- The rules for investing the account's assets (e.g. low risk overseas treasury instruments) must be clear, agreed and published.
- A high level oversight committee (key ministries plus qualified external advisers) must be established.
- There must be annual, professional audits of the oil reserve account.
- The oil reserve account, its management, and guidelines must be subject to rigorous transparency requirements.

In addition to savings and expenditure rules, the Government will have to choose investment rules for the saved funds. In determining these rules, it will be important to keep in mind that only investments matching the return on financial assets used to calculate the permanent expenditure per capita will make oil revenues sustainable. Saving the oil revenue abroad will shield the economy from the adverse effects of huge inflows of foreign exchange. Investments in infrastructure and human capital will have to be efficient and carefully managed in order to avoid adverse macroeconomic effects. A summary of the key elements that should be observed in setting up the rules governing the oil reserve account is presented in Box 3.6 below.

Current efforts and initiatives to improve the function of petroleum tax administration should also be deepened and strengthened. The authorities are conscious of the importance associated with the function of petroleum tax administration and have asked AUPEC to update and operationalize a highly sophisticated and detailed petroleum revenue model with this in mind. It would be difficult to overstate the importance of the petroleum tax administration functions in a country such as Angola. Beyond the obviously critical role it must play in ensuring that the State receives the revenues it is due, the tax administration function should be in a position to advise government on likely future tax or fiscal revenues for macroeconomics planning purposes.

A similar set of issues applies to the case of the diamond sector in Angola. The next Chapter provides an assessment of that sector, which alike the petroleum sector has traditionally been controlled by the Government. Yet, governance in the diamond sector has been considered significantly opaque, whereas the legal framework has been considered

> **Box 3.6: Elements of a Revenue Management Framework for Angola**
>
> **Revenue consolidation and collection**
> - The collection of all petroleum related revenues are consolidated through the oil reserve account
> - Revenues are accounted for according to agreed and transparent accounting guidelines
> - Revenues are published in an accessible and timely manner
>
> **Define savings and consumption**
> - Transfers from the oil reserve account to the budget are based on the agreed savings/expenditure rules.
> - The rules are clear, predictable, and public, do not depend on administrative or political discretion for their application
> - In the development of guidelines for savings and consumption, macroeconomic and sustainability concerns are paramount
>
> **Institutionalize the transfer mechanism**
> - Transfers from the revenue collecting authority to the budget and the oil reserve account follow predefined rules and occur automatically, independent of administrative or political discretion
>
> **Account management**
> - Management of the funds is based on clear, transparent, agreed, and predictable rules, which allocate clear responsibilities and reporting requirements.
> - The BNA is designated as the operational (day-to-day) manager for the account.
> - The rules for investing the account's assets must be clear, agreed and published.
> - Fund assets should to a large extent be invested abroad and in safe instruments
> - The BNA reports on performance of the oil reserve account and asset allocation according to a preset schedule, and the reports are made public in an accessible way
> - A high level oversight committee (key ministries plus qualified external advisers) must be established.
> - The oil reserve account, its management, and guidelines must be subject to rigorous transparency requirements

unstable. The potential of the sector to contribute more to the development of the country is huge, but more needs to be done so that this potential can be realized. The main challenges for the diamond sector are associated with problems related to governance, the business environment and the social contribution of diamond mining.

CHAPTER 4

The Diamond Sector: A Potential Underexploited

Angola is currently the fourth world's largest producer of diamonds with the potential to become one of the leading global producers. The industry has been traditionally controlled by the Government which has experienced difficulties to introduce in the diamond sector the same level of institutional and economic maturity observed in the oil sector. Partly due to the way the sector was affected by the civil war, frequent changes in the legal and regulatory framework concerned with the diamond industry and excessive government interference have prevented the major players in the global industry to contribute effectively to the sector's full growth. The main challenges that must be addressed by the authorities are related to problems with governance, the business environment, and the social contribution of diamond mining. This Chapter looks at these challenges and presents a reform strategy to unleash the potential of the diamond sector in Angola.

The Characteristics of the Diamond Sector

Angola became one of the world's major diamond producers in the 1970s, but quickly lost its dominance due to the independence war and a violent civil conflict. The start of diamond production in Angola dates back to 1912 in the Northern province of Lunda. This was and is the part of Angola where alluvial diamonds are most abundant. At that time, the sector was exploited by Diamang, a joint venture between De Beers, the Portuguese state and international mining finance interests, which was granted a monopoly over all diamond related activities in the country. Very quickly, diamonds became Angola's most valuable exports, and by 1971 Angola was producing 2.4 million carats a year, placing it among the world's major diamond producers. However, the war for independence and ensuing civil war severely disrupted mining and Angola dropped to

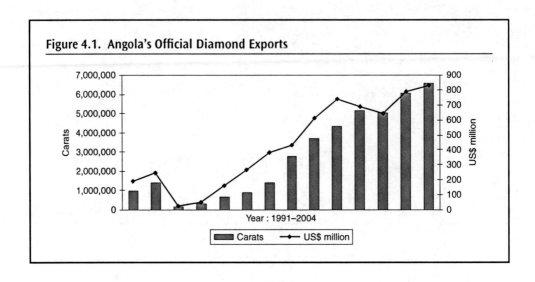

Figure 4.1. Angola's Official Diamond Exports

seventh place in world production with diamond output falling to 750,000 carats in 1975 and 350,000 in 1977.[41]

Many years of instability and the alluvial nature of production have contributed to the structural growth of illegal mining activities in the country. Illicit diamond production increased markedly in the late 1970s, resulting in the *Movimento Popular para a Libertação de Angola* (MPLA) government dividing the Lunda province into north and south sections in 1978 to restrict population movements. The MPLA also nationalized the control of the industry and, in the mid-1980s, a new state mining company, Endiama, took over Diamang's monopoly and operations with the purpose of revamping production. However, output continued to decline dropping to less than 100,000 carats by the end of the decade. The government liberalized the industry in the early 1990s in an effort to reverse the decline in production, but the main result of this policy was the rapid development of artisanal mining. The situation got worse during the civil war, when a large part of diamond production was in the hands of the rebel group UNITA. The result was a tremendous increase in illegal diamond output, reaching nearly US$600 million in 1992.

Most of the current formal and informal production comes from secondary alluvial deposits. Presently, the only kimberlite pipe exploited at commercial levels is Catoca, located in the Lunda Sul province. As of 2003, its output amounted to 3.2 million carats, equivalent to 65 percent of the total formal sector production by volume (see Figure 4.1). Catoca is currently in the process of doubling is production capacity, while a second kimberlite project, Luo, is expected to be commissioned in 2005. As a result of Catoca's expansion and the commissioning of new projects, Angola's formal exports in 2005 are expected to reach about 10 million carats, worth US$ 1 billion. The production of diamonds by project in 2003 is presented in Annex 2. In 2004, reported diamond exports totaled 6.63 million carats, worth US$ 763 million (see Annex 1). Angola's diamond

---

41. See Dietrich and Cilliers (2000), and Pearce (2004) for a detailed account of the history of diamond production in Angola.

exports represent about 95 percent of the country's non-oil exports and about 10 percent of non-oil GDP.

Currently, Angola is the world's fourth largest producer of rough diamonds in terms of value, with the potential to become one of the leading global diamond producers. The country has close to 12 percent of the share of the world market and a high proportion of its production is of gem quality. Diamond reserves were estimated in 2000 at 40 million carats in alluvial deposits, and 50 million carats in kimberlite pipes, which are just now beginning to be exploited. Around 700 kimberlites of varying sizes (10-190 hectares) and shapes are known in Angola, aligned along a SW to NE trend across the country and into the Democratic Republic of the Congo. But in order to unleash the full potential of the sector, underlying governance problems and the ensuing less than adequate business environment have to be tackled.

Governance: Opaque and Unstable Legislation

There is little competition and a large element of discretion involved in the concession of mining rights in Angola. The concession of mining licenses is granted as follows: (i) interested companies send to the Ministry of Geology and Mines a letter of intent explaining what sort of minerals they want to extract and outlining their technical and financial profiles; (ii) the Ministry evaluates the proposal and checks the availability of concessions and the final decision is made by the Minister; (iii) after approval of the concession, a contract is negotiated with Endiama; the Ministry charges a premium and a deposit which are also negotiable, and royalties which are payable once production starts; there is also an industrial tax of 35 percent on the net profit of the company; (iv) the maximum dimension of a concession is of 3,000 sq. km. A more competitive and transparent bidding system for new licenses should encourage all potential operators and generate an increase in state revenues.

The legal structure is not very clear about the roles of operator and regulating agencies in the sector. The Ministry of Geology and Mines is responsible for the implementation of the legal and regulatory framework for the sector, for issuing mineral rights, and for the geologic survey, while the mandate to approve kimberlite concessions is of the Council of Ministers. Endiama is by law the largest shareholder in all new diamond ventures and also has regulatory functions over the selection of companies that are to be granted new diamond mineral rights, the negotiation of mining contracts, and the monitoring and control of activities of diamond ventures. The company is therefore both an operator and a regulator in the diamond industry. In order to avoid potential conflicts of interests and to allow Endiama to exploit its commercial potential, its licensing, regulatory, marketing and advisory functions might be transferred to other agencies or to the current regulating ministry.

Due to developments associated with the lasting civil war, the government became at the same time a monopolist and a monopsonist in the sector. In 2000, Endiama transferred its exclusive marketing rights to a newly created subsidiary called *Sociedade de Comercialização de Diamantes* (SODIAM). Two basic reasons were put forward to justify the introduction of a monopoly in the commercialization of diamonds: (i) to improve collection of tax revenues from the companies operating in the sector; and (ii) to reduce smuggling and minimize the problem of the "conflict diamonds", as SODIAM would be issuing certificates of origins for the diamonds it traded, in conformity with the implementation of the Kimberly

> **Box 4.1: Certificates of Origin and the Kimberley Process**
>
> To limit the access of rebel groups to revenues from diamond sales, the Certificate of Origin Scheme was introduced in October 2000, in accordance with the United Nations Security Council Resolution 1306 (2000). The resolution imposed sanctions against rough diamonds imported from Angola and Sierra Leone without certification.
>
> The Kimberley Process was initiated in parallel, at a conference in South Africa in 2000, as a joint government, international diamond industry and civil society initiative to stem the flow of conflict diamonds. It has support from United Nations General Assembly Resolution 55/56. Participation is voluntary and includes governments of diamond trading countries, diamond trade organizations, companies with interests in diamond mining and trading, and NGOs with interest in issues on conflict diamonds. Angola is a full participant in the Kimberley Process. In November 2002, representatives from 40 countries and the European Union agreed to launch a certification scheme for rough diamonds in January 1, 2003. Kimberley Process Participants account for approximately 99.8 percent of the global production of rough diamonds.
>
> Under the terms of the certification scheme, each participant agrees to issue a certificate to accompany any rough diamonds being exported from its territory, certifying that the diamonds are conflict-free. Each country must therefore be able to track the diamonds being offered for export back to the place where they were mined (producing countries) or to the point or import (consuming countries). All importing countries agreed not to allow any rough diamonds into their territory without an approved certificate.
>
> Given the large volume of diamonds being traded across borders, a system of trade and production statistics was also introduced, in order to ensure that the volumes leaving one country match those entering another. This statistical system is still under development. Some countries, including Angola, have reportedly been particularly slow in disclosing their production and sales information.
>
> Trade monitoring and the lack of a system for regular independent monitoring of all national control systems remain important constraints for the effectiveness of the scheme. Furthermore, the whole process could be seriously undermined, if any one of the major diamond mining or related downstream manufacturing countries fails to participate effectively.

process (see Box 4.1). SODIAM was, however, faced with two issues at its inception: (i) its lack of comparative advantage in trading diamonds in key markets such as Antwerp; and (ii) its financial inability to deal with cyclical downturns in the diamond markets. To this end, SODIAM engaged in a joint venture with the Lev Leviev Group to form AScorp, a single official commercialization channel for all diamonds produced in Angola. The strategy was successful in bringing government revenues from diamonds up from US$10-12.5 million in 1998 to around US$739 million in 2000.[42]

The end of the war in 2002 brought yet another restructuring of the sector that established the end of the marketing monopoly and introduced different rules for transactions in the formal and in the artisanal markets. As the government regained control of all mining areas under rebel hands and formal production started to pick up again, Endiama announced the end of the marketing monopoly. AScorp's monopoly was terminated in 2003 and SODIAM was allowed to establish joint ventures with other companies to trade diamonds. In formal sector dealings, SODIAM determines the price of the diamonds after consulting

---

42. The arrangement was however harmful to long-term investments in the sector, insofar as ASCORP, given its dual monopsonistic/monopolistic position, could dictate artificially low prices on to rough diamond producers, thereby stimulating cross-border smuggling. The government estimates that at least US$ 1.4 billion worth of alluvial diamonds have been smuggled since 2000.

3 evaluators (1 presented by the producer; 1 presented by SODIAM; and 1 presented by the Ministry of Geology and Mines). Through an arbitrage process, an average and consensual price is fixed. SODIAM then imposes a discount on the average price. This discount is supposed to allow SODIAM to trade in the international market at competitive prices, but the practice has been criticized based on concerns about special and privileged treatments.

The implementation of the Kimberley Process has been relatively successful in Angola, but more could be done to improve the efficiency of the system. Angola was the first country to implement a full certificate of origin for diamond exports in 2000, and although implementation has been successful, a lot more could be done to improve the overall efficiency of the system. Apart from lack of field inspections by Kimberley Process officials, and some delays in reporting obligations, the most obvious problem is that there is no system to determine the origin of the diamonds from the informal sector beyond the records kept by the buying offices. To the extent that it is possible, it would be desirable to move toward the optimization and development of market efficiency and better incentives for a free market in the sector.

The Business Environment: Not Competitive Enough

The current set of diamond policies and institutions is producing mixed results in terms of the development of the country's huge diamond potential. The current business environment does not seem adequate to attract the type of investors that possess the financial resources and the technology needed to develop the existing kimberlite deposits. Large western mining corporations, when considering to invest in Angola, have expressed concern with relation to: (i) excessive Government discretion and lack of transparency on decision making, (ii) weak legal and regulatory framework, subject to fundamental abrupt changes in the rules of the game (namely in terms of ownership rights); (iii) poor definition of the key institutional roles and lack of powerful regulatory institutions; (iv) trade monopoly by SODIAM; (v) no clear rules for investment agreements with Endiama, with very strict articles of association (limited periods, no transferability, etc.); (vi) lack of accountability of Endiama management and poor disclosure of information; (vii) lack of security in mining areas; and (viii) lack of a clear framework for the sharing of revenues with the affected communities.

The performance of the fiscal revenues generated by the sector has been uneven. The fiscal contribution from the sector has improved dramatically since 2000, reaching US$ 44.6 million in 2002, and US$ 112 million in 2003, of which more than two thirds came from two projects (Catoca and SDM). However, preliminary figures for 2004 point to a sharp reduction in tax revenues to US$ 69.5 million (see Figure 4.2). From a fiscal point of view, taxes from the diamond sector have traditionally provided just 1–2 percent of total budgetary revenue.[43] The small contribution can be explained by the fact that the nature of

---

43. The logic for collecting taxes is as strong for diamonds as it is for tropical hard woods, but the most efficient way of doing it differs because diamonds are much easier to smuggle than logs. Export taxes on diamonds vary from 2.5% to 3.5% in most diamond producing countries. The proceeds lost by lowering export taxes could be offset by raising license fees, but the trade-off between the two would depend on local conditions (Goreux, 2001).

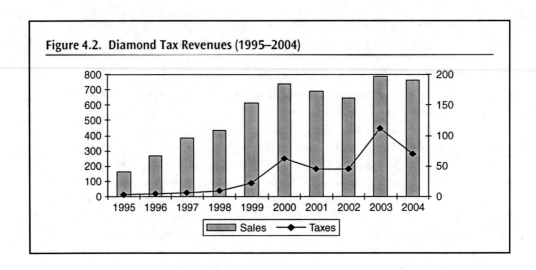

Figure 4.2. Diamond Tax Revenues (1995–2004)

the artisanal production of alluvial diamonds has traditionally made it very difficult to control production and trade, while facilitating smuggling activities.[44] On the large scale mining side, the government provides generous tax breaks and holidays to industrial projects in a proactive effort to share the potential benefits of the industry with the local private sector. Moreover, revenue from dividends of the government's participation in Endiama and the various diamond joint ventures are virtually nil.

The applicable fiscal regime for mining is not very complex. There are two main taxes: a corporate rate at 35 percent, and a capital income tax at 10 percent. Mining equipment and supplies are exempt from import duties. Fixed assets and exploration invested benefit from accelerated depreciation (50 percent in the year they are incurred, 30 percent in the second year, and 20 percent on the third). In addition, diamond production in Angola is also subject to three specific taxes:

- Taxes on surface rights of mineral permits are between US$ 1 and US$ 3 per hectare and per year and US$ 3 per hectare and per year upon renewal of the license.
- A bonus is payable on the award of mineral rights, based on the size and value of the project.
- Royalties for mining ventures are currently of 5 percent on the gross value of diamonds produced.
- Export duties are levied at a currently reported rate of 2.5 percent.

The fiscal burden on diamond production in Angola *per se* does not seem excessive when compared to other major producing countries, but the unstable and non-transparent legal

---

44. First, diamonds have very high value to weight and volume ratios facilitating their illegal trade. In addition, the mining sites are usually widespread and easy to work with simple technologies, making it difficult for Government to monitor the trade flows. Finally, the economic and social organization in the artisanal areas involves many actors, intricate divisions of labor, and informal business relationships.

framework reduces the sector's competitiveness. The limited transparency of the sector makes an assessment of the current taxation arrangements difficult. It is also difficult to assess how onerous or fairly applied is the current tax regime, as information on the margins earned by SODIAM and ASCORP and payments for mineral rights are not published. Moreover, Endiama takes a substantial free-carried interest in all ventures. Although these equity participations are seen by the foreign investors as a fiscal burden, reflected by the potential flow of payments to Endiama, they do not translate into fiscal revenues, neither in the form of dividends paid by Endiama, nor as signature bonus or any other form of quasi-fiscal revenue. Furthermore, the ambiguity and instability of the regulatory framework adds to the perception of Angola as a high-risk location and continues to deter investors. An international comparative assessment of the sector's fiscal regime is presented in Table 4.1.

Successful mineral development led by the private sector in Angola will require change. There are a lot of improvements possible in a sector that, to a large extent, remains secretive and victim of patronizing from the high instances of Government. Some of the critical issues involve: (i) continued peace and stability; (ii) stable macroeconomic environment; (iii) sound trade policies; (iv) predictable and transparent legal framework that adequately define investors' rights and obligations; (v) a fiscal package that is competitive and at the same time equitable for the concerned stakeholders; (vi) security of tenure of mining permits; (vii) strengthened capacity of the Government to monitor and regulate the sector; and (viii) a firm commitment towards marketing liberalization. The international experience suggests that there is scope for improvements in the legal framework currently in use in Angola (see Box 4.2).

A Three-Pronged Strategy to Unleash the Potential of The Sector

In order to improve the overall contribution of diamonds to the country, three sets of issues remain to be addressed:

- *Governance and Transparency:* Should Endiama keep both its functions of operator and regulator of the industry? What institutional reforms are needed both at Endiama and at the mining administration levels to improve governance in the sector? What else can be done to improve transparency in the sector?
- *Investment Climate:* How can the current investment climate be improved in order to attract and retain credible national and international investors to Angola's diamond sector?
- *Social Development:* How to increase the contribution of diamond mining to social development?

Improving Transparency and Governance

The diamond sector in Angola has been notoriously opaque undermining the full potential of the industry. Although Government intervention was instrumental to increase production and tax revenues from the sector, the overlapping of several functions inside Endiama—grantor of concessions, direct producer (both directly and indirectly, through its joints

Table 4.1. Fiscal Regimes for Diamond Mining

	Income Tax	Royalty	Import Duty	Dividend Tax	Export Tax	Government Equity
Angola	35%	5% of gross value	Exempt	10%	2.5%	20–49% free carrying interest
Botswana	25% to 51% (1)	10% of gross value	5% plus surcharge tax	15%	None	None, option of acquiring up to 15% fully paid equity
Central African Republic	30%	8% of gross value	Exempt		3%	None
Democratic Republic of the Congo	30%	4% of gross value	2% for equipment, 3% for inputs	10%	None	5% free carrying interest
Ghana	35%	3 to 12% of total revenue, based on profitability	Exemption	Exemption	None	None
Guinea	35%	5–10% of gross value for raw diamonds (2), 2% for cut diamonds	0–5.6%	15%	None	15% free carrying interest
Namibia	25 to 51% (1)	10% of gross value	Negotiated	10%	Negotiated	None
Sierra Keone	33%	3% of gross value	Exempt	None	License fees based on acreage	None
Tanzania	30%	3% of net back value	5% is the cap limit	10%	None	Negotiable
Zambia	35%	2% of gross value	Exempt	12.5%	None	Up to 30% fully paid equity
Brazil	15% on net profits and a surtax of 10% on any net profit exceeding US$ 100,000	Levied based on percentage of net invoice value	Mostly exempt or zero rated	None	Up to ten years income tax exemption but only in to northern areas of the country (restricted to some cases)	None

(1) With a variable tax formula based on the profitability ratio (taxable income as a percentage of gross revenue).
(2) Set according to profitability criteria defined by the Ministry of Mines.

Box 4.2: The International Experience on Regulating Diamond Production

Type of Contract	International Experience	Suggestion
Research and Prospection	There is no standard practice on this area	Parties should negotiate details of the contract regarding the work plan, calendars, budget, M&E systems, auditing and reporting
Exploitation	There is no standard practice on this area	The example from the petroleum industry should be used in the diamond industry because in the oil industry, these contracts are common and usually include the following: ■ Joint Operating Agreement ■ Project definition ■ Costing of the project ■ Calendars for operation and production ■ Terms and conditions for payment to partners ■ Sharing agreements ■ Operating committee ■ Reporting ■ Dispute resolution ■ Treatment of residuals
Royalties	There is no standard method to determine royalties	These are usually negotiated among the parties
Reserves	The diamond market is highly sensitive as to the definition of reserves and there are international rules on this fixed by the ISCs and their respective technical committees	A "standard" and acceptable model to define reserves should be adopted. Different rules should be used for the alluvial (USGS) and the kimberlite (OSC and AusIMM) deposits.

ventures), and exclusive marketer—imposes constraints to transparency, accountability, and impairs planning.

Regulatory and institutional reforms are needed, both at Endiama and at the mining administration levels. A priority task would be to enact transparent and stable legislation under a the new diamond law, providing a clear separation between regulators and operating companies, a simple and transparent management system of mining rights, ensured consistency in taxation and marketing arrangements, and defined objective criteria and minimum requirements for private investments, clearly indicating that special privileges would not be accorded to individual citizens or companies.

The authorities should reflect as to whether Endiama should keep both its functions of operator and regulator of the industry. Similarly to the case of the national oil company, discussed in the previous Chapter, there is a clear potential conflict of interest in the exercise of Endiama's operations as negotiator of every concession agreement, shareholder in every project, producer on its own account, and purchaser of all production from the country. To avoid conflicts of interest, and to permit Endiama to exploit its commercial potential, Government should consider transferring its licensing, regulatory, marketing, and advisory functions to the Ministry of Geology and Mines Mines (as it was the case before the position of Minister was given to an UNITA representative), while Endiama would keep its prominent position as a commercial enterprise.[45] Its role as a passive shareholder might also be separated from its active operational roles.

Financial oversight over the company should also be improved, with a more active role to be played by the Ministry of Finance. As a state-owned company, Endiama both generates and spends public funds on a massive scale. As several major projects are currently under consideration, investment requirements should increase considerably in the near future. These investments will have a substantial macroeconomic significance and need to be discussed in the context of a budget debate. The oversight over Endiama's operations requires an update of the first phase diagnostic of the diamond sector and financial audits of Endiama, carried out by PricewaterhouseCoopers for the five years prior to 2003. Both the diagnostic study and the audit reports should be published and made available to the public.

The updated diagnosis study should be followed by an effort to improve data collection, coordination, analysis, and decision making in the sector. This would involve the set up of an independent, transparent data collection system to monitor the sector, the improvement of data collection from artisanal areas, and of record keeping and information management across the sector. This could be a preparatory step for Angola to join the EITI for diamonds

Improving the Business Environment

Initial requisites to improve the business environment involve not only clearer rules for the sector but also better geological information. The long war meant that most of Angola has not been explored using modern techniques and only 40 percent of the country has been covered by basic geological surveys. The country is now considered to be one of the most promising diamond areas in the world with over 600 kimberlite pipes identified, most of which await more detailed geological evaluation. However, due to the marketing monopoly, lack of transparency in the grant of mineral rights, and overall country risk issues, few international world class investors have invested in the sector. This has delayed the development of several promising new projects.

The policy to share the benefits of the diamond sector with the domestic private sector also requires improvements. Under its current form, the policy could be compared to

---

45. The new diamond law is expected to clarify the relationship and the specific roles of the Ministry of Geology and Mines and Endiama.

a sale of "lottery tickets" to the private sector, providing them with an opportunity to try to attract foreign sponsors with the financial resources to develop these properties. Given the uncertainty of the investment climate and lack of reliable geological information, most of the foreign investors that accept to play the game are relatively small companies without the financial resources or the technical skills needed to develop efficient mining projects. They are attracted by the possibility of making quick gains by raising equity through private placements or in the international capital markets, to finance the minimum amount of work required to eventually "flip" these concessions to more reliable mining companies. Efforts to move towards the liberalization of the market should be considered along with the necessary legal and institutional reforms required for achieving this objective.

Despite the attractiveness of existing alluvial deposits, the cost structure of diamond production in Angola is still higher than in neighboring countries. Although the alluvial deposits represent easy targets for entering the industry, as their high grades, low-risk technical profile, and short commissioning period allow for almost immediate cash flow generation, the cost structure for diamond production in Angola is significantly higher than in neighboring countries like South Africa. Efforts to engage domestic private companies in the diamond sector are commendable but, given the low value added brought by the Angola counterpart to the venture, they represent in practical terms an additional level of taxation for foreign investors, refraining the development of the industry.

Strengthening the Social Contribution of Diamond Mining

The formalization and re-organization of artisanal and small-scale mining could be enhanced as a source of legitimate non-farm income generating opportunities, while promoting more socially and environmentally responsible mining practices. As it is practiced today, artisanal mining brings little economic benefit to the local communities. The bulk of the profits remains concentrated at the level of the traders while most of the diggers receive only a fraction of the sales price of the stones they extract. The "patrocinador", or middlemen, many of them from foreign countries, is often responsible for funding the operations, and providing rudimentary equipment, buying in exchange all the production from the diggers, mainly from Angola and DRC. Diggers are engaged on the basis of receiving subsistence from the artisanal mining license holder during the period of work. They are financially rewarded with fifty percent of any eventual diamond sales, over and above the working costs incurred by the license holder.

The authorities should consider adopting measures aimed at improving the overall standard of living of the communities linked to the mining activity. The government's current strategy to reduce smuggling of diamonds aims at curbing informal artisanal production by having the diggers absorbed by licensed companies that will use more efficient extraction methods, ensure safety standards, pay taxes, and provide social services to their local community. Most of the social initiatives in the diamond areas are left to the companies without a clear framework for the engagement of the mining communities and little support from the government (see Box 4.3). For this policy to succeed, the government should complement it with measures aiming to improve the overall standard of living of the involved communities, through the pursuit of multiple goals. A few priority areas to be considered by the authorities include: (i) improvement of artisanal mining rights and security of tenure; (ii) ensure that mineral wealth supports sustainable economic and social

Box 4.3: Corporate Social Responsibility in the Diamond Sector

Endiama and its associated joint venture companies have invested in 2003 about US$ 12.2 in community development, mainly through the contribution to the rehabilitation of the : (i) Malanje-N'zagi road (US$ 7 million); (ii) Provincial Hospital of Saurimo (US$ 2 million); (iii) schools in N'zagi, Cuango and Saurimo (US$ 0.7 million); (iv) Cafunfo's airport (US$ 0.2 million); and (v) humanitarian assistance (US$ 0.3 million).

Catoca has also invested in a number of social projects in its area of influence. The company has built a residential housing complex, as well as a primary school for 303 students, and provides meals for the students on an ongoing basis. It has also renovated the provincial hospital, built a potable water supply system for the Catoca surrounding areas that benefits some 6,000 residents, and sponsors sporting events.

Box 4.4: Time Frame and Actions for Diamond Sector Reform and Development

	Short-Term	Mid-Term	Long-Term
Large-scale mining	1. Update Diamonds Law in the context of international best practices in the area. 2. Based on this law, redefine the relations between the Ministry of Geology and Mines (as regulator), and Endiama (as operator). 3. Update the 2003 diagnostic of the diamond sector. 4. Disclose financial audits of Endiama. 5. Define frameworks for community investment agreements. 6. Liberalize diamond sales.	1. Enforce the newly revised frameworks. 2. Angola to join EITI initiative for diamonds. 3. Restructuring of Endiama. 4. Raise awareness and disseminate rules of the game locally and internationally. 5. Initiate compilation of geo-databases and aerial survey. 6. Define frameworks for community investment agreements.	1. Active Promotion of new projects in diamonds. 2. Disseminate basic promotional geo-information.
Small-scale and artisanal mining	1. Review and regularize existing mineral rights (exploration and mining) 2. Design program for the formalization of ASM and community development plans. 3. Reinforce implementation of the Certificate of Origin Scheme. 4. Strengthen enforcement of mineral rights. 5. Identify and assess options for free market trading of diamonds.	1. Implement program for the formalization of ASM, community development plans. 2. Identify and assess options for free market trading of artisanal diamonds. 3. Initiate pilot extension services and train field staff. 4. Implement pilot social plans for communities and disseminate best practices and results. 5. Disseminate geo-information.	1. Private based rent-lease of equipment. 2. Incorporate best practices and results of pilot social initiatives. 3. Pilot test selected options for free market trading of diamonds.

development of the communities; and (iii) providing a level playing field for the license holders as well as diggers through training to enhance their product evaluating capabilities.

Whatever the strategy selected by the authorities, an integrative approach should be pursued. Over the medium term, the interventions for improving the artisanal and small scale mining subsector should focus on: a) improving quality of life of those living and working in diamond fields; b) rehabilitating the environment and improving agricultural productivity; c) upgrading regional infrastructure; d) improving mining productivity and safety; and e) encouraging general economic growth via multiplier effect of mining activity and infrastructure improvement. Additional emphasis should also be given to: (i) continued implementation of the Kimberley Process; and (ii) the harmonization of the domestic policies with those of competing and neighboring countries (namely the Democratic Republic of the Congo), to discourage smuggling and encourage trading through official channels.

A phased reform strategy for the development of the diamond sector in Angola is summarized in Box 4.4.

Shared growth and sustainable development in Angola will demand a better investment climate and improvements in the business environment. Along with the petroleum and the agriculture sectors, the diamond sector hosts the hopes for sustainable development based on equitable and shared-growth in Angola. However, none of these sectors will meet the challenge of providing a better life to the Angolan people if the factors that inhibit private sector development and the sharing of the country's mineral wealth are not removed. As discussed in the previous chapter, in the case of the petroleum sector, the key issue is associated with the impact of the oil windfall revenues on macroeconomic stability and the quality of institutions. In the case of diamonds, there are equal constraints associated with governance, the business environment, and the potential of the sector to contribute for the increase of the welfare of the population. In the case of agriculture, to be discussed in depth in Chapter 6, the challenge is how to increase productivity in the presence of an overvalued real exchange rate. However, in order to create the conditions for the development of a thriving private sector that can interact with and benefit from the mineral sectors, a number of issues related with the quality of the business environment in Angola must be addressed. The next chapter addresses these issues.

CHAPTER 5

Private Sector Development and the Business Environment

While most of the world is suffering from some form of "petro pessimism," Angola is amongst a handful of countries that could derive benefits out of the current international scenario of high and rising oil prices. This raises the expectations on the speed with which the "peace dividend" could be shared with the Angolan population. However, a rapid economic recovery, with more jobs and income for the average Angolan, will be difficult to achieve without the necessary commitment to structural reforms. The climate for doing business in Angola—whether for residents or foreign companies—is perceived as one of the least conducive in the world and any strategy to promote private sector development in the country will need to rest on pro-business measures that can enable companies to compete more effectively in an open economy. This chapter discusses options and priorities to remove barriers and improve the investment climate in Angola.

The Private Sector in Angola

The Business Environment

The business environment in Angola is challenging. According to the *2006 Doing Business* survey of the World Bank, establishing a company in Angola takes an average of 146 days, more than twice the regional average. Licensing is a time-consuming and costly procedure. The time to comply with all licensing and permit requirements is estimated at 326 days, almost one year. Registering property also takes about 11 months and costs more than 11 percent of the property value. Investor protection is not high when compared to similar countries and obtaining credit is equally difficult. The average import requires 10 documents, 28 signatures, and 64 days. It is equally difficult to enforce a contract—in calendar days, it takes 1,011 days to resolve a dispute starting from the moment a plaintiff files a lawsuit in court until settlement or payment. Dispute resolution among Angola's neighbors and potential competitors takes much less time: in Zambia 274 days; 270 in Namibia; and 154 in Botswana.

Restrictive laws and business practices hamper the development of an appropriate investment climate. Angola ranks below many other Sub-Saharan countries on many measures of the ease of operating in the private sector, and ranks last out of 155 countries surveyed by the World Bank in its *2006 Doing Business* report. Cumbersome property registration procedures and a costly and insecure business environment are common features of the private sector in Angola. Other key investment climate constraints include: (i) labor market rigidities; (ii) time and cost to start up a business; (iii) access to and cost of finance; and (iv) contract enforcement. All of these constraints hamper entry and competition in the formal economy, encourage informality, and feed rent-seeking practices. The challenge facing Angola in creating an encouraging environment for private sector investment is summarized in Table 5.1 and reflects its poor scores on indicators describing the ease of doing business.

The challenging business environment creates additional costs, increases risks, and imposes barriers to competition. These are the factors that can influence profitability and job creation. The opportunities and incentives firms have to invest productively, create jobs, and expand can be traced through their impact on expected profitability. Furthermore, profitability is influenced by the costs, risks, and barriers to competition associated with particular opportunities.[46] Each factor matters independently, and all three are interrelated. Some risks can be mitigated, for example, by incurring greater costs. High costs or risks can be barriers to competition. And barriers to competition can reduce risks for some firms but deny opportunities and increase costs for others. Governments have more decisive influence over many aspects of the investment climate, such as the security of property rights, approaches to regulation and taxation, the adequacy of infrastructure, and the functioning of finance and labor markets. As outlined in Table 5.1, these are the areas in which Angola needs a dramatic improvement in order to create a better investment climate.

Besides the difficult business environment, corruption is often a reality in oil rich countries and Angola ranks amongst countries with the worst corruption perception indices. Oil revenues provide governments with a source of revenue independent of its citizens, diminishing the need for accountability, and diminishing the incentive to pursue the components of good governance listed above. Oil wealth may actually encourage the opposite behavior, and at the same time increase government's ability to buy off or intimidate opposition to such behavior. Those with ulterior motives of political or personal gain can be very successful at making the petroleum industry and its revenue flows opaque. In the rankings established by Transparency International's widely referenced Corruption Perceptions Index, petroleum exporting developing countries find themselves in the bottom one third of the countries listed. Angola's rank in the most recent release of the CPI was 151, number nine from the bottom. This is a particularly serious finding because corruption is recognized as one of the largest single inhibiting factors to private sector investment and growth.

The Quality of Infrastructure

In addition to corruption and an unfavorable business environment, another important inhibitor of private sector development in Angola is the lack of appropriate infrastructure.

---

46. These points are developed further in the World Bank's 2005 World Development Report, *A Better Investment Climate for Everyone*, Chapter 1.

Table 5.1. Government Policies and Behaviors and Investment Decisions

	Factors that Shape Opportunities and Incentives for Firms to Invest		
	Government Has Strong Influence	Government Has Less Influence	The Situation in Angola (Ranking out of 155 countries)
Costs	■ Corruption ■ Taxes ■ Regulatory burden ■ Infrastructure and finance costs ■ Labor market regulation	■ Market-determined prices of inputs ■ Distance to input and output markets ■ Economies of scale and scope associated with particular technologies	■ Corruption rank: 151 out of 155 countries (Transparency International) *Doing Business Indicators:* ■ Ease of Doing Business: 135 ■ Starting a Business: 155 ■ Hiring and Firing: 117 ■ Dealing with Licenses: 122 ■ Paying taxes: 79 ■ Getting credit: 77
Risks	■ Policy predictability and credibility ■ Macroeconomic stability ■ Rights to property ■ Contract enforcement ■ Expropriation	■ Consumer and competitor responses ■ External shocks ■ Natural disasters ■ Supplier reliability	■ Recent progress on macroeconomic stability *Doing Business Indicators:* ■ Registering Property: 145 ■ Enforcing Contracts: 122 ■ Protecting Investors: 48
Barriers to Competition	■ Regulatory barriers to entry and exit ■ Competition law and policy ■ Functioning finance markets ■ Infrastructure	■ Market size and distance to input and output markets ■ Economies of scale and scope in particular activities	■ It takes 146 days to start up a company and 335 days to register a property in Angola. ■ Commercial code dates back to 1888. ■ Infrastructure destroyed during war and still in the process of reconstruction.

Source: World Development Report 2005 (Table 1.1), *Doing Business 2006*.

Since independence, infrastructure provision has been almost a monopoly of the public sector be it in the form of Government departments, or of state-owned enterprises. With the long war and inefficient management of public companies, the country's infrastructure is in a perilous state, negatively impacting both the quality of life of the people and the operation of almost all economic activities. Many state industrial and manufacturing enterprises record substantial losses, which have in many cases depleted working capital. In comparison with the SADC countries, many of which are not highly developed, Angola scores at or close to the very bottom for most infrastructure indicators, such as access to safe water and sanitation, electricity, and telephones (see Table 5.2).

Table 5.2. SADC Infrastructure Indicators

Class of Human Development	HDI Rank	Country	Percent of Population with Access to Safe Water			Percent of Population with Access to Sanitation			Electricity kWh Per Capita Per Annum	Telephones Per 1,000 Population (2000)		
			Rural	Urban	Total	Rural	Urban	Total		Fixed Line	Cellular	Total
High (1)	47	Seychelles								235	320	555
Medium (7)	67	Mauritius								235	151	386
	107	South Africa	70	99	87	80	92	87	3,587	114	190	304
	122	Namibia	71	100	83	20	93	62	1,022	63	47	110
	125	Swaziland							576	32	33	65
	126	Botswana	88	100	90	41	91	55	755	93	123	216
	128	Zimbabwe	69	99	79	32	96	52	715	18	23	41
	132	Lesotho	57	91	62	35	56	38	188	10	10	20
Low (6)	151	Tanzania	58	92	66	83	98	86	45	5	5	10
	153	Zambia	10	81	38	57	94	71	609	8	9	17
	155	DRC	26	89	42	6	53	18				
	161	Angola	22	46	31	27	62	40	69	5	2	7
	163	Malawi	40	95	47	1	18	3	71	4	5	9
	170	Mozambique			63			54	40	4	2	6

Source: SADC, Compiled from World Bank (2005), Country Framework Report–Private Solutions for Infrastructure in Angola.

There are difficulties with infrastructure that are spread across different key sectors that range from transport to telecommunications. The road system is in a shocking state of disrepair, making several provinces all but inaccessible by road. In many areas, roads have been mined, and at least 300 bridges have been destroyed. The three railways systems at their height carried 9.3 million metric tons of freight to the Atlantic ports, but this has now fallen to insignificant levels. On the 1,340 km Benguela line, for example, services are now limited to the 30 km stretch between Lobito and Benguela, and it is mainly passengers rather than freight been carried (World Bank 2005c). Urban infrastructure also has dramatically deteriorated, the streets in most urban areas are in a state of virtual collapse, and there are essentially no functioning sewerage or drainage systems. The fixed-line telephone system is antiquated, very limited in coverage, and prone to service interruptions, while the two small cellular networks, largely servicing the capital, are unable to meet fast growing demand. Out of some 160 Angolan municipalities, as of 2003, only 40 were connected to the fixed public telephone network.

There have been efforts to repair the road network after the end of the war. The classified road network consists of 72,323 km of roads: 7,777 km of paved roads; 28,018 km of gravel roads; and 36,528 km of earth roads. Much of it has received little or no maintenance because many roads are in former war zones and have carried little or no traffic. In 2002 the government approved a program of emergency repair and reconstruction of roads. During the first phase, from 2002 until September 2004, basic repairs were to create the minimum conditions for traffic at a cost of US$55 million (US$45 million for roads and US$10 million for bridges). The second phase, which is underway now, envisages improving regional connections with the main roads and is budgeted at US$171 million. The General Government Program 2005–2006 emphasizes the needs of transport infrastructure, identifying 4,194 km of roads for rehabilitation, the construction of 2,000 m of bridges, and the rehabilitation of 17 bridges. Projects rehabilitating international road links to the Democratic Republic of Congo and Zambia were submitted to the COMESA Secretariat in 2003–2004 for feasibility studies. China also supports the rehabilitation of the road network. In addition, the World Bank has started an Emergency Multisector Recovery Project in six provinces, which includes a component devoted to roads.

The precarious road conditions inflate transportation costs which impact negatively on competitiveness. Though traffic has increased from the near-zero level of the war years, long-distance road transport is still an arduous, costly, and time-consuming undertaking. An 18-wheel truck that transports goods 1,250 km from Luanda to Dundo in the northeast of the country cannot make the round-trip in seven days, and will have no backload from Dundo to Luanda. The truck has to take 1,250 liters of fuel from Luanda because gas supply along the road is unreliable. It is not surprising that the cost of this type of transport is high: US$3,500 (down from US$5,500 in 2002/2003 because of heavy competition). A 423 km trip from Luanda to Malange costs US$2,500.

Inefficiency in service provision by state-owned enterprises is also widespread. At present, most infrastructure service providers are forced to make do with inadequate prices and therefore operate at a loss. Costs overwhelm revenues and enterprises depend on subsidies from central government. Subsidies generally cover varying proportions of both capital and operating expenses, are not targeted, and do not provide any sort of incentive to enterprise managers to improve efficiency. A recent household survey commissioned by the World Bank to investigate the incidence of subsidies and the satisfaction of the population with the public services showed a dismal picture, whereby large percentages of the population surveyed responded that the services provided are of minimal satisfaction. More than 50 percent of the households surveyed declared to be minimally satisfied with roads and sewage while some 30 percent of the respondents were minimally satisfied with schools, electricity provision, health center services, home tap water service and public transportation. In sum, relatively few households enjoy the key services provided by public authorities, even though, relatively many households agree that these services are of maximal importance.

The experience of involving the private sector in infrastructure rebuilding so far has been limited. As discussed in the World Bank's *Country Framework Report* of 2005, the most high profile of these is the mobile phone license issued to Unitel in April 2001. Telecommunications is also the only infrastructure sector for which a regulatory body has been established. In some other sectors there have been management contracts, which are considered to have been reasonably successful (water supply in Soyo and Caxito, solid waste collection in Luanda, terminal operations in the port of Luanda, and ground handling services in Luanda airport). New concession contracts are being formulated for solid waste collection and terminal operations in the port, while in the electricity sector a concession was recently awarded to Alrosa, and in the telecommunications sector four new fixed line licenses are being issued.

The Cost of Capital

Another constraint to private sector development lies in the financial sector. There is no stock market, the money market has very few tradable instruments, the insurance sector is dominated by two state companies, and there are no institutions specializing in housing finance. The banking sector consists of nine commercial banks, one merchant bank, four representative offices of foreign banks, and a few institutions providing loans to small-scale enterprises. Among the commercial banks, three are foreign-owned and two are state banks, BPC and BCI. The three larger banks (BPC, BFA, and BAI) handle some 82 percent of the turnover of the entire system. However, the three private banks recently established (BTA, BESA, and BCA) are the first to have offered banking products denominated in dollars, and have began to win a significant market share because of better quality service and a breakthrough in banking marketing. The state banks account for about 45 percent of commercial bank assets in the country. The Central Bank has recently created a division specialized in micro-finance in anticipation of an appropriate legal framework in response to growing demand by commercial banks to offer this service.

And as in many other African countries, the cost of capital in Angola is high. With a short term inter bank interest rate at 96 percent in 2003, loans made out to clients by commercial banks carried, on average, annual interest rates above 100 percent. The cost of capital is expected to remain high in the short term. The Economist Intelligence Unit estimates that short term interbank interest rate will average 94 percent and 96 percent in 2005 and 2006, respectively. Therefore, the cost of finance remains an important constraint to the development of the Angolan private sector and smaller businesses are likely to be constrained the most. Individually and combined, these factors have constrained the ability of businesses to undertake investments, including in technology and skills development, a factor which has hampered their productivity and competitiveness.

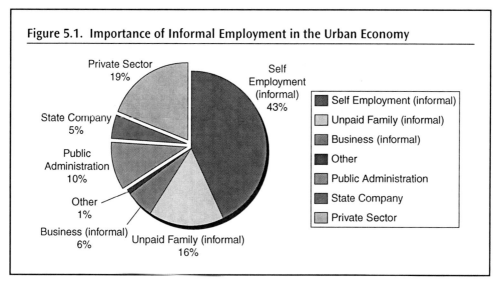

Figure 5.1. Importance of Informal Employment in the Urban Economy

Source: Cain (2004).

A Bimodal Private Sector

The private sector in Angola is bimodal and the two major poles are quite distinct from one another. On the one hand there is a thriving informal sector while on the other end there is a highly sophisticated sector linked to the oil economy. As portrayed by Grion (2004), in one extreme there is a segment of companies that operate under a legal and fiscal framework that is controlled by the state. These companies are usually foreign-owned, use sophisticated technology, and operate in competitive markets. In the other extreme, there is a myriad of micro enterprises that operate in a less dynamic market, using outdated technology that is intensive in unskilled labor, and trading mostly in the informal market. In the latter segment, the logic is more one of survival than entrepreneurship. The gap between the top and the bottom of this bimodal distribution—the so-called *missing middle*—is huge in Angola and there is scope for it to be filled with small and medium enterprises, which in the case of Angola have not yet reached a significant number. As shown in Figure 5.1, approximately 65 percent of business activities in urban areas in Angola are done through the informal sector. Self employment leads the informal categories, with 43 percent of the economic activity.[47]

The most dynamic pole of the economy is concentrated in Luanda. The only source of information on the private sector in Angola is a recent enterprise census carried out by INE in 2002 and made available publicly in 2004. Based on that census, there were 19,119 private

---

47. Research on the causes of growth of the informal sector points to excessive bureaucracy and government interventions in the market as two of the most important factors encouraging informality. Loayza, Oviedo, and Servén (2005), for example, argue that a heavier regulatory burden, particularly in product and labor markets, reduces growth and increases informality. The authors conclude, however, that these effects can be mitigated as the overall institutional framework improves.

Table 5.3. Trends in Merchandize Trade

	1995	1996	1997	1998	1999	2000	2001	2002	2003
					US $ (000)				
Total Exports	3753	4993	4768	3825	4829	7927	6881	8030	9486
Total Imports	1734	1930	2316	2047	2055	2023	3238	2886	4161
Trade Balance	2019	3063	2452	1778	2774	5904	3643	5144	5325
					Percent of GDP				
Total Exports	74.5	66.3	62.1	59.3	79.3	89.5	72.6	71.4	71.9
Total Imports	34.4	25.6	30.2	31.8	33.7	22.8	34.2	25.7	31.5
Trade Balance	40.1	40.7	31.9	27.5	45.6	66.7	38.4	45.7	40.4
					Composition of Exports (percent)				
Crude Oil	90.4	91.1	89.2	85.9	85.1	88.9	88.9	90.5	94.0
Refined Oil	3.1	2.2	1.5	2.0	1.9	2.1	1.7	1.7	1.9
Diamonds	4.4	5.1	6.9	9.2	11.5	7.7	8.2	6.3	3.2
Others	2.1	1.6	2.4	2.9	1.5	1.3	1.2	1.5	0.9
Total	100.0	100.0	100.0	100.0	100.0	100.0	100.0	100.0	100.0

sector firms employing some 340,000 people in 2002. Of those, 14,484 firms (75 percent) employ between 1–9 employees. The largest density of businesses is concentrated in Luanda, which abridges 55 percent of the total number of firms in the country, followed by the provinces of Benguela, Kwanza-Sul, Huíla, Cabinda, and Huambo which together account for only 26.6 percent of the total. The remaining 12 provinces host the other 30 percent of the existing companies in Angola, but they are virtually disconnected from the most dynamic markets of Luanda and Benguela by the virtual lack of any real linkages beyond these large economic poles.

Trade Patterns and Regional Integration

With several constraints to private sector participation in the economy, it is not surprising that Angola exports mostly oil and diamonds and imports almost everything. Angola's exports, 72 percent of its GDP, are heavily dependent on oil and diamonds, which in recent years have amounted to an average of about 93 percent and 6 percent of total merchandize exports, respectively (Table 5.3). Items such as stones, sand, fish, and so forth make up the remaining one percent. With limited refining capacity, almost all exported oil is crude oil.[48] On the other hand, Angola imports almost everything, including products that the country

---

48. Oil is exported largely under long-term contracts with particular countries. This is usually done with oil-backed loans which commit oil production to the repayment of the loans.

has a comparative advantage. Merchandize imports being about 30 percent of GDP, the trade account generates a large surplus averaging 40 percent of GDP. This is offset by payments for services related to investment in the oil sector and interest charges on large short-term external debt. The current account generates a deficit or surplus depending on the price of oil.

Tariff Structure

Angola applies MFN treatment to all its trading partners. Angola's tariff, like its import system overall, has been considerably revised and liberalized since May 1999, when a new tariff code (*Pauta Aduaneira*) based on the Harmonized System (1996 version) was introduced. At that time, the maximum duty rate was cut from 135 percent to 35 percent. The latest tariff schedule for imports and exports, based on HS 2002, with a maximum rate of 30 percent, was introduced in 2005 under Decree-Law No. 2/05.[49] According to the authorities, the objectives of the new tariff schedule are to:

- Align the tariff structure with that of the HS 2002, according to the requirements of the WTO and the World Customs Organization.
- Revise duties to protect national production without harming the interests of consumers and guarantee the supply of essential goods at competitive prices.
- Protect national industry from dumping practices; promote, as a strategy and with comparative advantage as a guide, a gradual process of substitution for imports of essential goods and to relaunch exports from non-oil sectors.
- Grant tariff benefits to the productive sectors of the economy and create fiscal equity.

The new tariff structure is oriented towards strengthening of the domestic production. Specifically, the new tariff structure seeks to reduce tariff rates that affect imports of raw materials; maintain or slightly increase tariffs on finished goods that can be acquired locally in acceptable quantity and quality; impose minimum rates on essential goods and intermediate products that are not produced locally or whose production levels do not satisfy local needs; and impose maximum rates on used goods not incorporated in local products.[50]

The tax system is straightforward, but tax rates are relatively high compared to other SADC countries. The principal legislation is the Law on Taxation Policy and Levels (Law No. 5/99). There is a series of other laws for specific taxes. Rates can only be changed by amendments being passed to the relevant laws. Corporate profits are taxed at 35 percent while profits on agriculture, forestry, or cattle-raising attract the only concessional corporate tax rate of 20 percent. Public enterprises are subject to taxation of profits at the corporate rate (35 percent) as well as other taxes. A tax on contracts is levied at the rate of 3.5 percent for construction, improvement, and repair of fixed assets and at 2.5 percent for all other contracts. Personal income tax rates on salaries and wages, bonuses, and benefits range from 4 to

---

49. The 2005 tariff is available at the following website: http://www.ita.doc.gov/td/tic/tariff/country_tariff_info.htm#Angola.
50. Information supplied by the Angolan authorities.

15 percent. Contributions to social security are compulsory and total 11 percent. The employee contribution is 3 percent and the employer contribution is 8 percent.

There are no quantitative restrictions on international trade and no incentives to exports. Licensing is required for all trade activities but they are granted almost automatically. Importation of certain products (transmitters, receivers, explosives, plants, fruits, seeds, drugs, etc.) require ministerial permit. In 2001, the Government introduced a Customs Expansion and Modernization Project to improve customs administration. The Project has been implemented in collaboration with the Crown Agent, a consulting company. Most products require pre-shipment inspection that is provided by BIVAC standards. There is no export incentives system in place (duty drawback, bonded warehousing, export processing zones) but investment goods in the oil and mining sectors are exempted from import duties.

Tariff Dispersion and Averages

Angola's applied tariff has six bands of 2, 5, 10, 15, 20, and 30 percent. There are no duty-free lines. The tariff follows the HS at the 8-digit level. Out of the 5,384 tariff lines in the schedule, 66 percent (3,570 lines) are rated at 2 or 5 percent; 38.5 percent (2,074 lines) are at 2 percent and classifiable as "nuisance rates," which are likely to cost more to raise than they yield. The two highest rates of 20 and 30 percent jointly apply to 10.5 percent of tariff lines, or 565 lines. The simple average MFN applied tariff in 2005 was 7.4 percent, by major sector (WTO definition), the average tariff on agricultural goods is 10 percent (compared with a bound average of 52.6 percent), and that on nonagricultural goods 6.9 percent (compared with 60.1 percent bound average—see Figure 5.2).

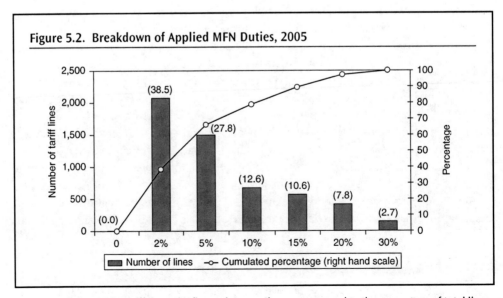

Figure 5.2. Breakdown of Applied MFN Duties, 2005

Note: Angola has 5,384 tariff lines; the figures in parentheses correspond to the percentage of total lines.
Source: WTO Secretariat calculations, based on data provided by the Angolan authorities.

Overall, the Angolan tariff is likely to provide domestic producers in selected sectors with high levels of effective protection.[51] Most tariffs on industrial inputs, capital goods, and equipment are low, or at "nuisance" levels (2 or 5 percent). In addition, substantial duty-free concessions are available to investors in priority zones, as well as to the oil and mining industries. The combination of low tariffs and concessions means that most investors pay little or no customs duty on inputs, equipment, and capital goods for at least the initial period (up to 10 years) of their activities, and in the oil and mining industries for the duration of their activities. At the other extreme, a number of sensitive domestic final goods, including certain construction goods, are taxed at relatively high nominal duty rates of 20 or 30 percent . Because of the low duties and tariff concessions on goods used in production, the effective protection of added value is, in these areas, likely to be many times higher than the nominal rates of duty on the final goods would imply.

Trade Partners

Angola participates in a number of trade agreements, but has not benefited from any preferential arrangements due to its lack of capacity to produce and lack of competitiveness. It is a member of the Southern African Development Community (SADC), but it does not participate in SADC's Trade Protocol. It also benefits from non-reciprocal preferential treatment from many industrialized countries under the Generalized System of Preferences (GSP) including the EU's Everything But Arms (EBA) Initiative and the US's Africa Growth and Opportunity Act (AGOA). Angola is a signatory to the Cotonou Agreement and, together with some SADC members, it is also negotiating a reciprocal Economic Partnership Agreement (EPA) with the EU which will replace the Cotonou agreement. Angola has not been able to take advantage of any of these preferential arrangements because of its lack of capacity to produce and its lack of competitiveness. It is essential therefore to remove the supply-side constraints to be able to fully benefit from these preferential arrangements, although this is likely to take some time.

Almost half of Angola's exports are shipped to the US and a quarter to China (all crude oil). China seems to be gaining some of the market share that until 1999 belonged to the US. The EU accounts for about 20 percent of total merchandize exports (Table 5.4). With regard to the sources of imports, the EU has the largest share at about 50 percent (Table 5.5), followed by South Africa (13 percent) and the USA (10 percent). Exports of both oil and diamonds are projected to increase substantially in the coming years. Such increase may not be translated into overall growth and poverty reduction unless the export revenue is used more productively and the policy environment is improved significantly to assist the economy to diversify.

Foreign direct investment in the oil and mining sectors is growing fast, but linkages with the rest of the economy have yet to be developed. In 2003, Angola was Africa's third largest recipient of foreign direct investment in Africa (with US$1,415 billion[52]), following Morocco and Equatorial Guinea. According to a recent report by the OECD,[53] three related

---

51. Effective protection is a measure of the protection provided to an industry by the entire structure of tariffs, taking into account the effects of duties on inputs as well as on outputs. See Corden, W. Max (1966).
52. See United Nations Conference on Trade and Development (UNCTAD, 2004).
53. See OECD (2005).

Table 5.4. Destination of Angolan Exports (percent)

	1995	1996	1997	1998	1999	2000	2001	2002	2003
European Union	21.6	19.3	14.3	17.0	17.0	17.7	26.3	26.4	13.4
France	2.4	2.9	3.8	2.9	2.0	4.7	9.7	7.9	7.3
North America	64.0	59.9	62.2	64.4	53.2	46.0	48.1	40.8	47.5
USA	63.7	57.2	62.2	64.1	53.2	45.6	47.6	40.8	47.5
Latin America	3.0	5.3	3.8	2.3	1.5	2.4	3.6	1.4	2.4
Brazil	1.1	3.1	0.8	0.6	0.6	0.4	2.8	0.2	0.1
Asia	10.8	12.9	18.6	15.9	27.3	33.3	21.3	30.5	36.0
China	3.6	4.9	12.7	4.0	7.4	20.8	10.5	13.5	23.3
Sub-Saharan Africa	0.5	1.4	1.3	0.3	0.9	0.6	0.8	0.9	0.7
South Africa	0.0	1.1	0.9	0.1	0.7	0.5	0.6	0.5	0.6
Others	0.0	1.1	0.9	0.1	0.7	0.5	0.6	0.5	0.6
Total	100.0	100.0	100.0	100.0	100.0	100.0	100.0	100.0	100.0

phenomena—the discovery of new oil fields, the increasing cost-effectiveness of deepwater exploration in a context of high oil prices, and the strategic interest of American business and non-traditional OECD partners such as China and India in the energy potential of the South Atlantic—are driving FDI activity. Chevron Texaco, in particular, has earmarked $11 billion for investment over the next five years. Despite their positive contribution to GDP and exports, oil projects have very high import content and very few

Table 5.5. Sources of Imports (percent)

	1995	1996	1997	1998	1999	2000	2001	2002	2003
European Union	60.4	50.0	29.1	31.8	44.9	46.2	37.9	45.8	51.6
Portugal	19.8	20.6	19.9	20.0	14.3	16.8	13.8	18.7	17.7
North America	15.2	14.1	12.4	17.6	12.7	11.6	8.6	13.1	12.2
USA	15.0	13.9	12.1	17.3	12.3	10.8	8.5	12.9	11.8
Latin America	2.3	3.9	3.7	6.7	3.5	5.9	5.0	8.2	7.4
Brazil	1.2	1.8	3.5	5.9	3.1	5.2	4.4	6.9	5.6
Asia	9.4	9.0	19.9	9.2	19.5	9.5	28.0	8.7	11.1
China	1.2	1.5	1.3	1.8	0.8	1.7	1.4	2.1	3.5
Sub-Saharan Africa	10.7	21.6	6.1	7.8	12.3	20.8	16.2	18.7	14.9
South Africa	8.6	19.7	4.7	6.1	10.3	16.8	12.9	15.7	13.2
Others	2.0	1.5	28.8	26.9	7.2	6.0	4.2	5.4	2.8
Total	100.0	100.0	100.0	100.0	100.0	100.0	100.0	100.0	100.0

Source: IMF Direction of Trade.

linkages with local business. Although the number of backward and forward linkages has started to grow (foreign companies have ad hoc programs to increase local content) the integration between the domestic and foreign businesses remains limited to very low-skilled activities such as catering and cleaning services. The rest of the economy, however, continues to attract very little FDI due to several constraints to the business environment.

Strengthening of the non-mineral private sector and diversification of the economy will require improvements in the legal and regulatory framework. Recent improvements in trade volumes, private sector development, and rising FDI levels are encouraging, but there is an urgent need for deeper and further reform in the legal and regulatory framework. All of these examples are important in demonstrating that the private sector can thrive in Angola, but in view of the long history of centralized control of the economy, a number of constraints to private sector development have yet to be removed. A number of such constraints is discussed in the following section.

Recent Actions to Improve the Business Environment

Many of the problems faced by businesses outside the mineral economy in Angola are associated with the level of development of its institutions. As argued in previous Chapters, the appreciation of the real exchange rate and other phenomena associated with the Paradox of Plenty are per se hard to deal with when the objective is to foster broad-based growth, but recent research on the drivers of growth and economic development reveals that institutions also matter a lot.[54] Broadly speaking, institutions include the various checks and balances that underpin democracy, the laws of contract and property, and the regulatory mechanisms that underpin the market economy. In a country with weak institutions, the absence of checks and balances leaves corruption unrestrained, and corruption impedes economic development by generating inefficiencies that block entry and competition. Angola performs poorly in an international comparison of institutional quality amongst oil-rich countries and has therefore a daunting challenge to promote growth outside the mineral economy (see Figure 5.3).

There have been encouraging initial steps to streamline and reform the regulatory framework. To consolidate its pro-market economic policy stance, the government committed itself in 2002–03 to overhaul the legislative framework for private sector activity in Angola, which was heavily based on Portuguese commercial codes that date back to 1888. Of the nine main laws, supplemented by various decrees, which constitute the overall legal framework for private investment in infrastructure in Angola, one was revised in 2002 and four are new legislative instruments, which were passed in 2003. These are the Law on the Delimitation of the Sectors of Economic Activity (revised in May 2002), the Private Investment Law, the National Private Investment Agency Law, the Law on Tax and Customs Incentives for Private Investment, and the Voluntary Arbitration Law (all passed in April 2003). In addition, a Land Tenure Law was passed in 2004 with the aim of clarifying property rights and customary tenure.

---

54. See Collier and Hoeffler (2005); Hausmann and Velasco (2005); Fosu O'Connell (2005); Rodrick, Subramanian and Trebbi (2002); and Rodrick (1999).

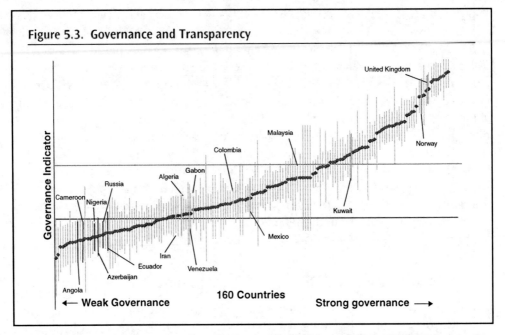

Source for data: http://www.worldbank.org/wbi/governance/govdata2001.htm. This chart shows estimates of control of corruption for 160 countries during 2000/01, with selected countries indicated for illustrative purposes. The vertical bars show the likely range of Governance indicators, and the midpoints of these bars show the most likely value for each country. The length of these ranges varies with the amount of information available for each country. Colors are assigned according to the following criteria: Red, less than 30% of overall countries rank worse; Yellow, between 30% and 70%; Green, over 70%. Countries' relative positions in no way reflect the official views of the World Bank or the International Monetary Fund.

Despite these promising efforts there are still significant risks for investors. The main investor risk arises from the provision in the sectoral laws which enable the state to nationalize assets when it deems it in the nation's interests to do so. However, the Constitution, the Private Investment Law and the sectoral framework laws do guarantee the foreign investor the right to compensation if this occurs, in accordance with international law and international dispute resolution procedures. At this point, there is also a high level of regulatory risk in that regulatory frameworks are largely underdeveloped. The only stand-alone infrastructure regulatory body presently in operation (telecommunications sector) does not have an adequate degree of autonomy from government. Without a more robust legal framework for infrastructure regulation, there is a high risk of political intervention in the granting of licenses and concessions and the determination of prices.

There have been encouraging measures to improve access to credit, but results have yet to materialize. A recent report by the OECD and the AfDB quotes that *Novobanco,* a micro-finance bank also active in other Southern African countries, has developed financial instruments and a system of credit lines that bypass red tape hampering access to finance for established businesses. According to the report, in the three months since its opening in Luanda in September 2004, the bank had already extended more than 120 credit lines in a total amount of $600,000 (the average loan was for $5,000, maturing in 3–5 years, at a monthly interest rate of 4 percent), almost entirely to clients operating in the trade sector. Such a successful uptake was made possible by the flexible formula offered to small entrepreneurs, which includes a no-fees account with no minimum balance, informal guarantees (house assets and a guarantor) and an ongoing relationship with loan officers. A network of such officers is responsible for assessing portfolio quality and monitoring clients, for which they are paid performance-related salaries. The report notes, however, that the USAID-financed scheme lacks a technical assistance component, which is considered one of the main requirements for small business development.[55]

Other equally encouraging initiatives have been launched to foster private sector development. As previously discussed, these include a new investment law that provides equal treatment to foreign and Angolan firms (with exceptions); a new commercial code enacted in early 2004 to replace the 1888 commercial code and the 1901 law on limited-liability companies; the establishment of the National Private Investment Agency (ANIP); and the creation of a one-stop registration office for companies (the so-called *Guichê Único*). One of constraints that this new office faces is concerned with delays in the publication of new firms by the *Diário da República* due to lack of representatives from the national official press in the *Guichê Único*. Additional provisions will be required before the new commercial code becomes fully effective while the one-stop registration office has yet to show practical results.

There have been positive initiatives to reduce barriers to competition, but these are still strong. The telecommunications sector is the only sector for which a regulatory body has been established and significant challenges remain to be addressed on that front. Despite the move towards liberalization, additional progress can be made in introducing competition by removing barriers to entry and in particular by licensing additional mobile service providers. Angola Telecom remains a dominant monopoly provider in many areas, as new entrants who could make a difference face in practice numerous obstacles such as the non–transparency of tariffs or the inability to obtain timely and cost-effective interconnection. These obstacles further result in lack of competition, high prices, and a communications infrastructure limited mainly to businesses and government, and to major urban centers. Additionally, there are some constraints in the legal and regulatory framework. There remain overlapping responsibility areas between INACOM, the regulator, and the Ministry of Post and Telecommunications, and the regulator has only a limited measure of autonomy. As of 2003 private investment was not authorized in the basic network infrastructure and it is therefore unclear which services are left to the newly licensed public fixed line operators. The telecommunications sector in Angola is hindered by the lack of a clear, effective and transparent regulatory framework. INACOMfurther faces difficulties with lack of skilled staff.

A number of privatizations have taken place, but the privatization program needs to be revamped. In 1989, a public agency, GARE (*Gabinete de Redimensionamento Empresarial/ Cabinet of Enterprise Redimensioning*), was created to coordinate and execute the privatization

---

55. The OECD/AfDB report also highlights that, recently, other initiatives have combined lines of credit to small businesses with training and technical assistance. In particular, a local bank, Banco Sol, has gradually started financing individual businesses from its traditional clientele, requiring informal collateral and relying on international NGOs for monitoring and assistance to the clients.

program of State companies. The program was divided into two phases. Phase 1 was implemented in the period of 1990–2000 and accounted for the restructuring and privatization (global and partial) of 409 companies yielding some US$100 million to the Government. In 2001, the government launched phase 2 of the privatization program covering the period 2001–05. This program involved about 90 public companies, most of them being from the previous program. Those companies are part of the following sectors: fishing, trade, hotels and tourism, transport, agriculture, manufacturing, oil, geology and mining, electricity and water, communication, and finance. Official data indicates that the biggest volume of revenue in this phase was reached in 2002, when privatization had raised about US$14 million. In the first six months of the current year, the State collected US$2.5 million from the privatization of eight companies. Total revenues of US$20.5 million have been collected since the second phase of the privatization program was launched in 2001. By the privatization of four beer companies (CUCA, NOCAL, N'GOLA and EKA) is expected to be completed before the end of 2006.

The creation of a national development bank of Angola will not help private sector development if the business environment does not change. The authorities have announced the establishment of a national development bank which will direct some 5 percent of the government's oil revenues for subsidized lending to the private sector without collateral or an adequate equity stake. Several countries have tried the same route and failed, including Angola in the past. The international evidence suggests that such institutions promote inefficiency and moral hazard and that they are prone to governance problems, including elite capture of subsidized capital and the emergence of non-performing loans. More effective ways to improve access to finance for micro, small and medium-size businesses include microcredit, venture capital, and supportive action on contract enforcement. But none of these options will be successful in promoting private sector development if the overall business environment in Angola does not improve soon.

The Way Forward

There is a world of opportunities for investment in Angola both in the mineral and non-mineral economies that can be realized by removing barriers and constraints to the investment climate. As described in Chapter 1, Angola is blessed with natural resources, profuse climatic diversity, and a hibernating market economy. With the advent of peace since 2002 the country now faces the daunting challenge of channeling its huge resource endowment into reconstruction of its infrastructure and into poverty reduction activities. However, in the aftermath of the civil war, improvements in the competitiveness of the local industry and efforts to diversify the economy away from the mineral sectors remain hampered by inadequate infrastructure, poor governance indicators and a less than adequate business environment. In what follows, a few suggestions on how to address these issues are discussed.

A Phased Approach to Improve the Business Environment: What Should be Done in the Short Term

A first step to improve the business environment is the definition of a clear set of priorities. Improvements in the investment climate in Angola will require reduction in unjustified costs, risks, and barriers to competition. In practice, costs, risks, and barriers to competition are a function of government policies and behaviors that play out through a wide range of specific policy areas. But where should Angola begin? No reform of any kind can be successful if a clear and consistent list of priorities is defined. In defining the priorities for reform, the authorities need to consider the current conditions in the country, the potential benefits from improvement, the links with broader national and regional goals, and implementation constraints. A good way of starting this priority setting exercise in Angola could be through the structuring of a dialogue mechanism between the public and private sectors. The establishment of a consultation mechanism between the government and the private sector is a critical element to secure a private-sector-led growth agenda.

Effective prioritization and sequencing can be done through formal consultations with the private sector. To accomplish this, it is first necessary to institutionalize such mechanism and the first step in that direction could be the realization of an annual private sector conference. To be effective, this meeting would require the presence of senior Cabinet members. In some countries, such as Senegal, Malaysia, Mozambique, and Zambia to name a few, the Head of State presides over the meeting. The private sector and the authorities should then determine together the sequencing and prioritization of the necessary reforms. In that instance, starting up with "quick wins", such as the adoption of the reduction in the time and cost for the registration of a business,[56] necessary adjustments in labor market legislation, among others, seem to be warranted.[57]

In parallel to consultation with the private sector, the authorities can also carry out an inventory of the regulatory environment in Angola in the very short term. The authorities should carry out a systematic inventory of existing rules and regulations which affect the country's investment climate. This should not only include an analysis of general restrictions for how to setup a business, but also sectoral analysis. Such a review could be done by an independent and professional team of consultants, with the participation of the private sector. The objective of the comprehensive review is to identify the existing investment climate constraints in a systematic manner. Once these constraints are identified, a time-bound action plan to revoke and/or simplify rules and regulations that cannot be justified or are cumbersome would be drawn up. Initiatives to that effective could be taken also in the very short term. The World Bank could assist the authorities on this regard through its *Investment Climate Assessment* (ICA) report.

Another short-term action is concerned with the reduction of the time required to register businesses in Angola. In order to address this issue, it will also be necessary to streamline the requirements for business registration and integrate the operations of all institutions involved in this process. The three institutions involved, namely the Notary Office, the Public Commercial Registry (*Conservatória de Registos Comercial*) and the State's Printing Office, (*Imprensa Nacional*) should be merged into a joint Management

---

56. For example, there is a need to reduce the minimum capital requirement for the establishment of a limited liability company which is currently set at $20,000 and the requirement for a down payment of at least $6,000 that must be paid in before an enterprise is established.

57. Additional details regarding the structure and operation of this mechanism could be developed through a specific study on this topic, which would also be based on additional consultation with stakeholders.

Information System (MIS) so that the relevant public officials from any one of them can retrieve the information entered by another institution, process it and make it available for processing by yet another institution involved in the registration. Under such an MIS, the Notary Public would prepare and make available the public deed of incorporation in an electronic version. Moreover, on the same day, the *Conservatória de Registo Comercial* will be able to record extracts of the public deed of incorporation and register the company. The same electronic document would then be made available to *Imprensa Nacional* for publication, while other necessary steps, such as the various inspections of the enterprise, are being carried out. Naturally, incorporation of business would need to be centralized at the country level so as to avoid different registries issuing the same name for different enterprises.

What Should be Done in the Medium and Long Terms

In the medium term, reforms that can enhance the potential benefits of changes that can be introduced in the short term should be favored. The impact of any policy improvement will depend on how it addresses a constraint that is actually binding on firms. For example, clarifying rights to land can help ease access to credit by firms and households, but only when it is relatively straightforward to start a business and register property. Another example is that reducing barriers to competition will not deliver its full potential if weak bankruptcy laws slow the exit of less efficient firms, or if labor market policies limit the ability of firms to adjust production processes to respond to a more competitive environment. Similarly, efforts to encourage local R&D can be hobbled by shortages of skilled workers, limited competition, or weak intellectual property rights.

Improvements in trade facilitation should ease some of the constraints to the investment climate. An efficient transport system, customs administration, and health and safety standards infrastructure are essential to developing a cost-effective trade network. The Government has already initiated steps to reform customs administration. The program should include upgrading equipment and infrastructure, integrating border agencies, reducing border clearance time, and training staff. Angola will also need to develop a standards infrastructure and testing services to be able to export agricultural products in the future. As a country with a lengthy coastal line and a number of neighboring land-locked countries, Angola has the potential to provide port and transit services to its neighbors. To realize this potential, it is necessary to develop a transit strategy and to harmonize customs procedures in collaboration with neighboring countries and SADC.

To complement the reforms associated with the regulatory burden in Angola, restrictions to property and land registration should be eased. As a first step, it is critical that land registration be facilitated. As mentioned above, it currently takes 335 days to register property with a cost equivalent to 11 percent of the property's cost. The Angolan authorities could facilitate the formal registration of land by reducing both the time and cost required to register it. Additionally, the government could scale up the allocation of land to private investors, who could then develop the area. This approach has the dual advantage of generating the funds necessary for infrastructure development and decreasing the price of formal land. Relaxing tenant laws, zoning restrictions and building codes is also a relatively easy and quick way to increase formal land availability.

It will also be necessary to revise labor regulations to introduce dynamism in the labor market. In Angola, when an employer decides for operational reasons to dismiss an employee, actually doing so requires a high degree of persistence and patience. The terms and conditions for employment termination are costly, time consuming and cumbersome. It is important to note that from an investor's perspective a barrier to firing is also a barrier to hiring. Therefore, it is important that the cost of dismissal be reduced, but this has to be done while still ensuring the protection of workers' rights. A revision of the Labor Law should be guided by the principle of adequately balancing the flexibility of firms to adjust their work forces in response to market forces and the need to protect workers' rights. It is equally important to ensure that the amount of severance payments and the limitations on working hours need to be brought in line with Angola's competitors and international best practice.

To realize the potential benefits of policy reforms, the authorities need to introduce measures that can catalyze changes. Change tends to occur when something shifts the incentives for maintaining the status quo. International experience illustrates how a diverse range of factors can trigger policy change even in the face of resistance by beneficiaries of the status quo. Those triggers can include external shocks and crises, technological change, new opportunities, new information and institutional competition, political change, and the initiative of policy entrepreneurs. In Angola, the potential benefits of policy reforms can only be realized if the bottlenecks to improved productivity and productive capacity in the nontradable sector of the economy can be removed. These bottlenecks are usually associated with the Paradox of Plenty and are concerned with physical and human capital limitations that constrain the expansion of the supply of nontradables.

To promote the expansion of the supply of nontradables and thus act in a catalyst way, it will be necessary to strengthen governance institutions, invest in physical and human capital development. The government is already promoting investments in roads, ports, telecommunications, energy transmission, and training of skilled workers. In what concerns the development of human capital, the authorities should consider ways to attract skilled diaspora Angolans. As discussed earlier, the lack of human capital is a severe constraint, especially in the public sector. At the same time, there is a large pool of highly educated Angolans outside the country. A major obstacle to their return is that the cost of an equivalent lifestyle is so much higher in Angola than in Europe or in the Americas. As oil wealth grows, one possible use for it would be to subsidize the return of key diaspora Angolans, perhaps through upfront grants for housing and education. There would obviously be costs in terms of divisiveness, but similar exercises have been successful in other countries such as Guinea-Bissau, Afghanistan, and Georgia, for example. In addition, it is important to be aware that improvements in the business environment will bring about the destruction of some jobs and the creation of new ones. This will require compensatory policies that target the most vulnerable groups with social programs (see Chapter 7 for a proposed approach).

The authorities also need to establish a long-term strategy to take advantage of trade agreements. Angola needs to determine its long-term trade interests and participate actively in regional and global trade negotiations to ensure that its interests are adequately reflected in future outcomes. Regional markets are important for Angola for non-mineral exports. It is advisable that Angola explores the terms upon which it might enter the SADC FTA. Maintaining market access to the EU market does not require Angola to sign an EPA unless it offers benefits beyond those provided by the EBA Initiative. To be beneficial to Angola, it is essential that an eventual EPA contains (a) improvements in the EBA and the

Cotonou rules of origin (RoO), (b) increased financial assistance to address supply side bottlenecks and to offset the revenue loss from lower tariffs on imports from the EU, and (c) adequate flexibility in EPA design to accommodate differing conditions among the countries within the SADC group. The Cotonou Agreement recognizes the need for developmental

---

Box 5.1: A Scorecard for Tackling Governance and Corruption

The challenge of governance and anticorruption confronting the world today strongly argues against the "business-as-usual" modus operandi. A bolder approach is needed. A common fallacy is to focus solely on the failings of the public sector. The reality is much more complex, since powerful private interests often exert undue influence in shaping public policy, institutions, and state legislation. In extreme cases, "oligarchs" capture state institutions. And many multinational corporations still bribe abroad, undermining public governance in emerging economies. There are also weaknesses in the nongovernmental sector. Further, traditional public sector management interventions have not worked because they have focused on technocratic "fixes," often done through technical assistance importing hardware, organizational templates, and experts from rich countries. Given the long list of interventions that have not worked, as well as the role often ascribed to historical and cultural factors in explaining governance, it is easy to fall into the pessimist camp. That would be a mistake. First, historical and cultural factors are far from deterministic—witness, for instance, the diverging paths in terms of governance of neighboring countries in the Southern Cone of Latin America, the Korean peninsula, the transition economies of Eastern Europe, and in Southern Africa. Second, there are strategies that offer particular promise. The coupling of progress on improving voice and participation—including through freedom of expression and women's rights—with transparency reforms can be particularly effective. For that reason, at the World Bank Institute we have begun to construct an index to help make transparency more transparent. Further, in terms of reforms, a basic checklist, which countries may use for self-assessment, includes:

- public disclosure of assets and incomes of candidates running for public office, public officials, politicians, legislators, judges, and their dependents;
- public disclosure of political campaign contributions by individuals and firms, and of campaign expenditures;
- public disclosure of all parliamentary votes, draft legislation, and parliamentary debates;
- effective implementation of conflict of interest laws, separating business, politics, legislation, and public service, and adoption of a law governing lobbying;
- publicly blacklisting firms that have been shown to bribe in public procurement (as done by the World Bank); and "publish-what-you-pay" by multinationals working in extractive industries;
- effective implementation of freedom of information laws, with easy access for all to government information;
- freedom of the media (including the Internet);
- fiscal and public financial transparency of central and local budgets, adoption of the IMF's Reports on Standards and Codes framework for fiscal transparency, detailed government reporting of payments from multinationals in extractive industries, and open meetings involving the country's citizens;
- disclosure of actual ownership structure and financial status of domestic banks;
- transparent (web-based) competitive procurement;
- country governance and anticorruption diagnostics and public expenditure tracking surveys (such as those supported by the World Bank); and
- transparency programs at the city (and subnational) levels, including budgetary disclosure and open meetings.

Source: Kaufmann (2005), *10 Myths About Governance and Corruption,* Finance and Development, IMF, September.

support from the EU to the signatories. Therefore, a successfully negotiated EPA would enable Angola to receive financial support from the EU to improve the production capacity. When the production capacity improves, Angola would be in a position to export particularly, agricultural commodities, to the neighboring SADC members duty- and quota-free under the SADC Trade Protocol.

Improvements in governance and anti-corruption actions are also required to unleash the full potential of private-sector led growth. Governance and corruption remain a sensitive and misunderstood topic that is usually avoided in policy dialogues, but it is not by avoiding the problem that it will go away. Recent research has extensively examined the impact of governance and corruption on development. The empirical results generally show that countries can derive a very large "development dividend" from better governance. The World Bank estimates that a country that improves its governance from a relatively low level to an average level could almost triple the income per capita of its population in the long term, and similarly reduce infant mortality and illiteracy (Bellver and Kaufmann 2005; Kaufmann, Kraay, and Mastruzzi 2005). Governance also matters for a country's competitiveness and for income distribution. In the case of corruption, research suggests that it is equivalent to a major tax on foreign investors. Yet, care must exercised when tackling corruption and governance, as it is not by the incessant drafting of new laws, decrees, and codes of conduct that results will be generated—what is most important is the need for fundamental and systemic governance reforms (see Box 5.1). In Angola, continued commitment to improve governance and corruption indicators is necessary since, as described previously, the country is ranked amongst those in the bottom end of corruption perception indices published by different sources.

Addressing the issues discussed in this chapter will be key to unleash the potential of the agricultural sector. The next chapter investigates the main challenges and opportunities currently found in the agricultural sector in Angola. The analysis builds on the recommendations of this chapter and focuses on ways to improve the productivity in agriculture and on the need to support smallholders.

CHAPTER 6

Removing Obstacles to Agriculture and Rural Development

Angola is a natural food and cash crop producer where agriculture can play an important role to support growth outside the mineral economy, provided that macroeconomic issues and agricultural policies are appropriately addressed. On the macroeconomic side, there is a high risk that the impact of an overvalued exchange rate can rapidly undermine agricultural competitiveness while at the same time budget execution rates for the agricultural sector remain extremely low. A realistic vision of the future for Angolan agriculture should include a mix of commercial and familiar farms. In this context, public agricultural policies should primarily focus on the development of smallholders but, at the same time, foster a conducive environment to encourage investments in the private commercial sector. This chapter revisits these issues while also addressing the need to promote institutional reforms, including the strengthening of recent decentralization efforts, the revision of investment allocations and the role of the Ministry of Agriculture and Rural Development (MINADER) in the development of the agricultural sector.

A Sector Facing Daunting Challenges

In line with the current reality in Africa, the large majority of the poor in Angola lives in rural areas. In the whole of Africa, roughly 80 percent of the poor are in rural areas, and even those who are not will depend heavily on increasing agricultural productivity to lift them out poverty. As producers, 70 percent of all Africans (and nearly 90 percent of their poor) work primarily in Agriculture. As consumers, all of Africa's poor (both rural and urban) count heavily on the efficiency of the continent's farmers, since farm productivity and production costs are fundamental determinants of the prices of basic foodstuffs which account for 60–70 percent of total consumption expenditure by low income groups.[58] In

---

58. See Gabre-Madhin and Haggblade (2003), and World Bank (2000).

Angola, between 60 and 70 percent of Angolans earn their living from agriculture, which currently accounts for less than 10 percent of GDP. According to the *Inquerito aos Agregados Familiares sobre Despesas e Receitas* (2000/2001), 94 percent of rural households were poor, compared with 57 percent in urban areas, due to the isolation from essential services and markets, and the destruction of their crops and livestock.

Agroclimactic conditions in Angola vary widely and there are large sparsely populated areas which are underutilized. The central highlands contain large areas with good rainfall (1500–2000 mm/year), relatively moderate temperatures, and are also the region with the highest population density. It is estimated that more than two thirds of the rural population live in this area. Soils, though adequate in some micro regions, are generally depleted of at least some macronutrients and require fertilization for sustained cultivation. The coastal and southern areas are far dryer, with average rainfall of less than 100 mm/year in southwestern desert areas and between 500 and 1000 mm/year in the four southern provinces of Namibe, Huila, Cunene and Cuando Cubango. Irrigation is essential to production in these zones, and fortunately there are abundant surface and subsurface sources, many of which have been developed to some degree. While population is densest in the high potential areas of the central plateau (Bie, Huambo, Kuanza Norte, Malange, Uige and parts of Benguela and Kuanza Norte), there are large sparsely populated areas, especially in the east, which are capable of supporting much larger populations than at the present time.

The successful performance as an exporter of agricultural products vanished with the war. The country was once the world's third largest exporter coffee. Maize was also a major export in the 1970's, amounting to more than 400,000 MT in its peak year, almost all of which was produced by smallholders. Cotton, sugar cane, sisal, bananas and wood also used to be important cash exportable crops. After independence, most of the settlers left the country and many of the former commercial farms and plantations were converted into state farms, which have now been privatized. A large number of rural inhabitants either fled or reverted to subsistence production. Moreover, infrastructure suffered greatly with widespread destruction of roads, bridges and warehouses together with the presence of thousands of land mines in rural areas resulting in a virtual collapse of the marketed production.

The country is well endowed with agricultural resources which remain mostly untapped. Staple crops range from *cassava* in the humid north and northeast to *maize* in the central highlands and sorghum/millet in the dryer southern provinces. *Potatoes* are an important crop in the central plateau and *rice* is also grown over large areas in the north. *Cattle* is raised over broad areas in the central plateau but are particularly important in the southern provinces of Cunene, Huila, and Namibe where there are an estimated 3 million heads of cattle. *Coffee*, the most important cash crop during colonial times, grows well in the highlands from Uige and Malange through Kuanza Norte and as far south as Huambo and Bie.

Recent increases in agricultural production are encouraging, but yields still are remarkably low. Cassava has experienced the largest increase in production since the cease fire agreement and the current production amounts up to 6 million tons/year. In the case of maize, production has reached pre-war levels over the last 2 years, averaging 600 thousand tons/year. Coffee production, however, is reduced to one tenth when compared to levels registered 30 years ago (see Figure 6.1). Accounting for population growth over the last 40 years, there has been a reduction of roughly 60 percent in per capita terms in the

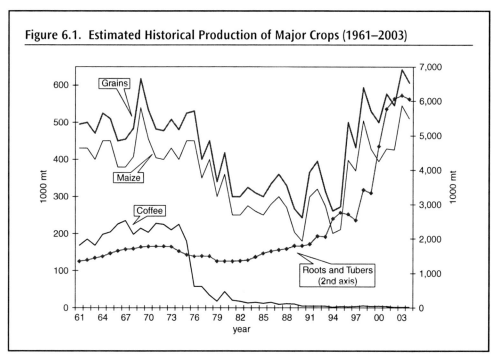

Figure 6.1. Estimated Historical Production of Major Crops (1961–2003)

Source: FAOSTAT.

production of cereals whose consumption has been partially substituted by roots and tubers, whose production per capita has increased by more than 70 percent (see Table 6.1). Furthermore, Angola still presents remarkably low yields when compared to other African countries. The only exception is cassava, whose yield is very close to the average (see Table 6.2). Although there is substantial room for improvement a doubling of yields would still leave Angola in the lower ranks of the range of yields found in comparator countries. In the authorities' view, large investments to improve the quality of the soil will be required before large scale agricultural production can contribute again in a significant way to the country's economic output.

Removing Obstacles to Output Growth

Reduce Costs

The inevitable overvaluation of the real exchange rate in Angola will remain as a key constraint to the competitiveness of agricultural products. The typical overvaluation of the currency in real terms observed in oil-rich countries can be a major disaster for the agricultural sector. In Angola, for example, every farmer wishing to produce maize for the coastal urban market has been effectively taxed by the real appreciation of the Kwanza, especially over the past 2 years. Casual evidence indicates that they may not be competitive with imports at current exchange rates, but that they enjoy some de facto protection in the interior due to the extremely poor roads. *It is thus clear that once the logistical infrastructure is rehabilitated, the rural farm production will have to compete with cheap imported*

Table 6.1. Production of Selected Farm Products (1961–2003)

Year	Popn (000)	Maize Total (ton)	Maize Per Capita (kg)	All Cereals Total (ton)	All Cereals Per Capita (kg)	Roots and Tubers Total (ton)	Roots and Tubers Per Capita (kg)	Coffee Total (ton)	Coffee Per Capita (kg)
1961	4,890	430,000.00	879.00	540,000.00	1,112.00	1,355,000.00	2,771.00	168,000.00	344.00
1965	5,180	450,000.00	869.00	562,000.00	1,085.00	1,580,000.00	3,050.00	205,000.00	396.00
1970	5,588	456,000.00	816.00	576,000.00	1,031.00	1,781,000.00	3,187.00	204,000.00	365.00
1975	6,187	450,000.00	727.00	558,000.00	902.00	1,542,000.00	2,492.00	180,000.00	291.00
1980	7,048	360,000.00	511.00	435,000.00	617.00	1,355,000.00	1,923.00	43,260.00	61.00
1985	8,299	250,000.00	301.00	308,000.00	371.00	1,580,000.00	1,904.00	12,000.00	14.00
1990	9,340	180,000.00	193.00	248,500.00	266.00	1,799,000.00	1,926.00	5,000.00	5.00
1995	10,868	211,000.00	194.00	296,000.00	272.00	2,762,500.00	2,542.00	3,300.00	3.00
2000	12,386	394,000.00	318.00	519,900.00	420.00	4,683,619.00	3,781.00	4,260.00	3.00
2002	13,184	430,000.00	326.00	550,000.00	417.00	5,782,000.00	4,386.00	2,160.00	2.00
2003	13,580	430,000.00	317.00	550,000.00	405.00	5,782,000.00	4,285.00	2,160.00	2.00
2003*	13,580	545,146.00	401.00	670,244.00	494.00	6,439,639.00	4,782.00		

* FAO/WFP Crop and Food Supply Assessment Mission to Angola, June 2003.
Source: FAOSTAT on-line data as of April 2004.

Table 6.2. Comparative Yields for Key Crops, 2003			
Crop	African Developing[a] (kg/ha)	Angola[b] (kg/ha)	Comparative Performance (%)
Maize	1 423	550	39
Cassava	8 835	8 850	100
Millet	703	350	50
Groundnut	855	390	46
Beans (dry)	608	220	36
Coffee (green)	448	19	4

[a] FAOSTAT April 2004.
[b] FAO/WFP Crop and Food Supply Assessment Mission to Angola, 2003 (except coffee, from FAOSTAT).

goods and tradable commodities, such as cereals. In this case, a possible way to prevent wide spread depression in the agricultural sector is by rapidly increasing its competitiveness.

In the face of an overvalued real exchange rate, the country should explore possibilities for reducing agricultural unit costs. The country has a better rainfall than many of its neighbors, has substantially lower yields, and has been cut off from technological advances in new varieties or other areas for decades. Cutting costs may take time and so further tariff reductions for agricultural trade should be implemented only as productivity improvements are realized. The key question, however, is whether productivity gains can be large enough to offset the disadvantages posed by the strong currency and high transportation costs. As there is no clear indication as to what is the actual cost structure of production in Angola, this question can only be answered through a case by case accurate crop budget analysis.

The recent experience of Nigeria can be relevant for Angola. Due to competition from artificially cheap food imports, made possible by the appreciation of the real exchange rate, producers in the major agricultural regions could buy imported food at the farmgate at prices below their own production costs. The result was a collapse in domestic production and a conversion of what had been one of Africa's largest agricultural exporters into one of its largest importers. Another important consequence of the oil boom and the ensuing overvaluation of the domestic currency in Nigeria was that the agricultural GDP fell on a higher level than that observed for agricultural employment while more than 60 percent of the country's arable land remained idle.[59]

Increase the Importance of Agriculture in the Budget

Increasing the priority given to agriculture and rural development through budget allocation will be an important measure to strengthen the sector. Budget allocations to the Ministry

---

59. The Nigeria case has been exhaustively studied both within the World Bank and outside of it. Major studies include those by Gelb et al. and Lele et al. in the 1980's and Schiff and Valdes in the 1990's, which document the evolution of the real exchange rate and its effects on agricultural production in the interior of the country.

of Agriculture and Rural Development (MINADER) in 2004 reached Kz2.3 billion (approximately US$25 million). This represents only 0.64 percent of the national total budget. Capital allocations to the agricultural sector have been somewhat better, at just under 2 percent of the Public Investment Programme (PIP), or Kz1.5 billion. Some Kz5.9 million were also budgeted for irrigation development in the separate 2004 Infrastructure Rehabilitation Programme (PRINF) controlled by the Ministry of Public Works. It should be noted that a considerable portion of the PIP (over 40 percent) was allocated to provincial governments in 2004. Of this amount, an analysis of provincial expenditure budgets indicates that a total of 8.2 percent (approximately Kz4 billion) were assigned to agriculture and other rural purposes; 2.3 percent (Kz1.65 billion) were purely for agriculture, mostly in the form of irrigation rehabilitation. Non-irrigation infrastructure absorbed the majority of provincial capital expenditures.

However, budget allocations alone do not suffice, as it is the actual budget execution that counts—and in this area serious improvements are needed in Angola. A common feature of the budget cycle in Angola is that the inclusion of expenditures in the State budget does not automatically guarantee that the funds are made available to the Ministry in question. Specific expenditure authorization has to be obtained from the Ministry of Finance and this is granted in tranches during the financial year, with a percentage held back for contingencies. Therefore, it is not necessarily the case that all budgeted expenditures are, in fact incurred. According to MINADER officials a significant, but not quantified, proportion of the budget has not been authorized over the last years, thus reducing expenditures below the budget figures. As a result, almost the entire disbursed budget is destined to personnel costs, penalizing current costs, such as fuel for the extension workers to visit farmers.

Move Gradually Towards Greater Decentralization

Much of the institutional apparatus of the Angolan government is unchanged from its command-economy structure of the past and needs to be revised. Decentralization of services, through the support of rehabilitation and capacity building at the provincial and local levels is critical. Despite the approval of a National Strategy for Decentralization in 2001, many aspects of the proposed decentralization policy remain undecided and it is still unclear which level of Government will have authority over different activities, and from where they will obtain the necessary financial resources. The absence of a clear decentralization policy has increased the difficulty of rehabilitating and maintaining rural infrastructure and services. In particular, extension services are located at the municipality level and it is there that assistance should have the greatest impact on target populations.

Decentralization should be encouraged as agriculture is perhaps the prototypical example of an activity that is *not* best managed from a capital city. While there is merit in keeping some activities managed and funded centrally agriculture is by its very nature a dispersed and decentralized activity. This fact, together with Angola's extremely wide variation in agroclimatic zones, infrastructure, and market proximity make linkage of the strategy for agriculture and the strategy for decentralization a high priority. If the key to long term growth is productivity improvement at the farm level coupled with development of marketing chains to connect farmers with demand centers and/or ports, there are necessarily many expenditures which must be made at a level well below that of the country or

> **Box 6.1: Decentralization Matters: The Mozambican "Proagri"**
>
> 1. While it would be a mistake to think that Angolan issues are directly parallel to those faced in Mozambique, there is at least one important lesson that can be learned from the post-conflict experience in that country–that is decentralization matters. As in Angola, the long conflict led to a virtual disappearance of government support services from many rural areas, due both to security issues and to the extremely low levels of funding allocated to agriculture.
>
> 2. The Objective: Rehabilitation of agriculture and of agricultural support services in Mozambique were identified as a priority area for donor support. Major donors, including the World Bank, supported the formation of a unified sector investment plan called Proagri through which all support was channeled, and which was administered by a joint steering committee including donors and representatives of the agricultural ministry. In this manner it was thought that the ministry capacity to manage project implementation and service provision could be strengthened and improved at the same time that badly needed farm level assistance could be provided.
>
> 3. The Lesson: Now entering into its second phase, the available evidence suggests that the channeling of all aid through the central ministerial apparatus has in practice made of the ministry itself the primary beneficiary of the project. The result was that less of the aid reached directly the farm level where it was most needed. Indeed, several observers of the Proagri have noted that the strengthening of the central bureaucracy, though necessary, has not resulted in any concrete benefits for many farmers and less benefits than hoped for at even intermediate levels (provincial, districtal, etc.). Remedying this perceived deficiency is one of the primary goals of the second phase of the project.

even the province. Box 6.1 presents some of the lessons that could be drawn from the Proagri in Mozambique.

The move towards expenditure decentralization, however, should be conducted gradually to avoid waste and inefficiency. Equally important concerns exist with regard to the capacity of local government structures to establish and operate the necessary administrative and budgetary mechanisms that decentralization would require. This is especially the case for municipalities which currently have no budgets at all (all existing local resources are managed from the provincial level), nor any revenue collection mechanisms except those managed by the Ministry of Finance.

Develop and Support a Market-Oriented Environment

There should also be a shift in policy towards further market deregulation. Implementing the ECP will involve a distinct change of focus away from humanitarian aid and resettlement/rehabilitation towards more development-oriented interventions. In the short term, this would represent adjusting and/or phasing out food aid programs to avoid unfair competition with the emerging domestic production of cereals. The lack of adjustment on such programs has been an important deterrent to domestic production in some countries facing post-conflict situations. An illustrative example of this market distortion was identified in the beginning of the 1990's in Mozambique. It was noted that local prices for maize were highly correlated with shipments of food aid in the post-conflict period, demonstrating the ability of large food aid shipments to depress local prices, and penalizing domestic production (Dunovan 1996).

Existing marketing controls need to be reviewed. In the area of marketing and distribution the government has long ago ceased to set prices centrally, but it is still in the business of setting wholesale and retail margins. These regulations are stated as limits of 25 percent on each of three market levels: producer or importer to wholesaler, wholesaler to retailer, and retailer to consumer. Though these laws are only sporadically enforced, they remain a distinct barrier to progress in the area of marketing. The standard prescription for exploitative marketing and trading (if it is occurring) is some combination of two strategies. One is to promote the entry of more traders into the market to provide competition. The other is to provide farmers with some measure of countervailing power through the development of producer associations. Government controls have demonstrated in Angola and in many other countries that they result in a poorly developed trading network and poorly serviced farmers. See Box 6.2 for a glance at the case of Brazil.

The government should not engage in public production or market activities. While the government has divested itself of many of its large state run operations in the sector, there is still a tendency to look to the government for things which in the past were done by parastatal organizations. Perhaps the most obvious of these is the state agricultural machinery company, MECANAGRO, which is now re-extending itself into rural areas after decades of virtual total absence. Recently, MECANAGRO has started to offer subsidized sales of seed and other inputs, undercutting possible private sector development. A standard rule of thumb would be that the government would not be responsible for any activities inside the farmgate, nor would it be responsible for any marketing services on the input or output side.[60]

There is scope to revise the current policy of distribution and subsidized sales of fertilizers. The distribution of subsidized fertilizer by the government is extremely detrimental to market development since a viable fertilizer supply system will depend on demand from commercial farmers as well as small farmers to be viable. As an interim measure to deal with the market's inability to overcome risks associated with input markets, the Government could auction donated or other fertilizer imports in provincial capitals. In the longer term, consider guarantees to support financing of private sector imports and consider the feasibility of domestic production. To stimulate demand, implementation of a targeted fertilizer voucher program similar to that used in Malawi could also be considered. This has the virtues of both stimulating demand and reducing risk for the private sector while keeping the government out of the business of direct distribution. This is an option that should be studied extremely carefully before going ahead given the very large optimal scale of fertilizer plants and the less than stellar record of such facilities in developing countries. Nevertheless, the ready availability of raw materials in Angola makes it at least a viable candidate for a feasibility study.

---

60. Another example is INCER (National Institute of Cereals) that remains the owner of various warehouses in varying states of disrepair in grain producing zones and along transport routes. The example of Mozambique is useful in this regard, where the government grain board rented or sold such facilities to the private sector in order to promote private sector marketing development. Clarification of INCER's mission in the future would be a useful step.

Box 6.2: Shrinking State Intervention and Soaring Productivity—The Case of Brazil

The agricultural and agroindustrial sector in Brazil has been one of the most spectacular examples of economic success over the past decades, at the same time that public intervention shrunk. While the public support, mainly through credit, has sharply declined over the past two decades, the grain production has nearly tripled (graph below). Currently, Brazil is the world's largest exporter of soybean, sugar, beef, coffee, orange juice, and tobacco; second largest exporter in soybean meal, chicken, and soybean oil, and fourth in pork, corn, and cotton. In contrast, currently, Brazil provides relatively very little public support to its farmers, according to OECD estimates (2005). Producer support in Brazil accounted for 3 percent of the gross value of farm receipts in 2002–2004–a rate far below the United States (17 percent), Mexico (21 percent), European Union (34 percent) and the OECD average (30 percent).

Evolution of the formal credit (in US$ million) and crop production (in million ton) in Brazil (1975–2005).

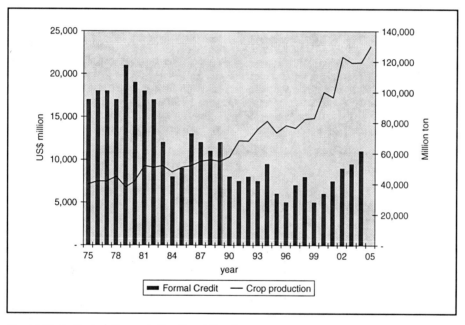

Source: Banco Central, Banco do Brasil and Conab.

A variety of factors contribute to this performance such as remarkable productivity gains associated with the incorporation of technological development, investments in infrastructure, and the deregulation and the opening of commerce. Other aspects not directly linked to the sector, such as economic stabilization (through the "Real" Plan, in 1994), fiscal adjustment of the country, foreign exchange policy, greater liberalization of commerce and a regional integration process are also pointed out as being fundamental for the sustainability of this process.

In terms of agricultural policy, the country has experienced a significant transformation that began in the seventies. Until that time, the policy model was based on a highly protected economy, substitution of imports, abundant subsidized supply of credit, and used floor price guarantees. The fiscal insolvency of the State and the economic instability that marked the eighties, however, lead to a near collapse of the rural credit policy at the beginning of the nineties. The failure of that model of credit, associated with the rapid and unplanned economic opening of the nineties, culminated in a major crisis in the agriculture sector.

(continued)

> **Box 6.2: Shrinking State Intervention and Soaring Productivity—The Case of Brazil** (*Continued*)
>
> With the crisis, a new and very competitive model has emerged: (i) government started to concentrate its efforts on the support to small holders and create stable and transparent rules for its interventions to avoid market distortions and discouragement of the private sector; (ii) government has created new instruments to support the sector, acting on a more localized and transparent manner, for example the launching of put options—a typical private sector tool-, in order to support producers in certain regions. (iii) as a consequence, the private sector came to the game and started to play a fundamental role on the provision of credit and risk sharing–currently, the private sector, such as processors, traders and input companies, are providing up to $5 billion dollars per year in credit for producers, through both cash and input.
>
> In addition to the need to continue improvements in agricultural competitiveness, Brazil, as many other developing countries, also faces a number of social challenges associated with agricultural development. A great deal of effort still has to be put forth to target small farmers and the rural poor for public assistance. Agricultural transformation is moving rapidly towards more competitiveness, and small family farms still represent 4 million rural properties (around 85 percent of the total), and 30 percent of the total planted area. Currently, poverty rates are influenced by two competing forces. On the one hand, economic growth at the national levels helps raise incomes, and generates demand-linkages throughout the economy. On the other hand, structural changes pose a threat to poor producers who are progressively less able to compete. Income growth in the Centre-West (the most competitive agricultural region) has been strong enough to reduce rural poverty, even though inequality has increased. Rural poverty has fallen more slowly in the North-East and actually risen in the North (where the rural population has actually grown), meaning that rural poverty is increasingly located in the poorest and less agricultural competitive regions

Improve the Effectiveness of MINADER

In order to function properly and effectively contribute to the development of the agricultural sector, it will be necessary to promote performance-enhancing changes in MINADER. Currently, MINADER is responsible for agriculture, irrigation, forestry, food security, issuance of rural land titles (although it has recently lost its role in cadastral and land registration to the Ministry of Urban Affairs), agricultural research and agricultural extension. To improve its effectiveness the Ministry would benefit from:

- *Performing an overall manpower study of MINADER* in order to identify redundant or unproductive positions and formulate a plan for their elimination.
- *Reforming the structure of MINADER* to strengthen the extension and research services, and pursue a unified management with the explicit mission of improving conditions of smallholder agriculture. This could be accomplished through a joint research/extension committee with oversight responsibility staffed with representatives from the research institutes, the extension system, the university faculty of agriculture, and farmer associations.
- *Creating a high level policy analysis unit* in MINADER capable of analyzing policy alternatives and appraising projects according to commonly accepted standards. This would require, at least, (i) adequate training and education for existing staff; and (ii) recruitment of new staff trained in economics.

- *Improving the existing statistical unit* within MINADER that would liaise with the National Institute of Statistics and the Ministry of Planning to design a national agricultural survey to establish baseline data regarding areas planted and amounts harvested (to be carried out by INE).

Stimulating Competitiveness through Better Incentives

Improve the Regulatory Environment

The agricultural sector in Angola faces several constraints to competitiveness. A scorecard was prepared by World Bank staff based on existing information about four main categories of factors that affect competitiveness in agriculture: (i) macroeconomic environment; (ii) structural issues; (iii) natural resource endowments; and (iv) vulnerability and risk. The full details of the scorecard are attached in the annex to this chapter. The scorecard findings indicate great potentiality in expanding agricultural activities once structural issues concerned with the quality of infrastructure and institutions are resolved. Major challenges involve the management of the real exchange rate and improvements in the allocation of funds and better budget execution rates for the agricultural sector.

The development of the sector will require investments in infrastructure and a better regulatory environment. A realistic vision of the future for Angolan agriculture should include a mix of commercial and familiar farms. This is an important debate in the country and it should influence the types of approaches and the priorities adopted in channeling support to the agricultural sector. Both need improved transport infrastructure, a modern marketing system and a conducive regulatory environment. The large smallholder sector in Angola is accustomed to producing for the market and has demonstrated a large capacity of responsiveness to market incentives, which can be seen as a natural comparative advantage in relation to other sectors.

In order to build on comparative advantages, obstacles to the development of the smallholder sector should be removed. There are excellent reasons to emphasize assistance to smallholders. The main reason is that these households, constituting a majority of the Angolan population, are less able to make needed investments than larger farmers. There are also compelling reasons at both the micro and macro levels for such an emphasis. At the micro level the need for an emphasis on aid to smallholders can be summarized as economic efficiency, poverty reduction, social cohesion, and political reconciliation. At the macro level, such an effort is critically important for the following reasons: potential food deficits (about half of Angola's grain needs are currently met through imports, much of these free or concessional), long term development, and impact of oil economy.

Rehabilitate Rural-Urban Commercial Circuits

A key goal to be pursued is the reactivation of rural-urban commercial circuits which can generate a self-sustaining growth dynamics. Rural producers, both large and small, need assistance in realizing the country's underlying comparative advantage in agricultural production. Producing for already existing food markets in cities as well as supplying newly reactivated manufacturing will ensure integration of rural areas into the economy on the output side. On the input side agricultural development will increasingly rely on purchased inputs while rising incomes will generate demand for manufactured consumer goods. The most striking gap in many areas is the near total destruction of rural marketing institutions and infrastructure. Without outlets for their produce, smallholder development at the farm level will be for naught. Among the most pressing needs for promotion the smallholder development are:

- *Road, bridge and rail rehabilitation and demining:* Many villages live in near total isolation. Improving market access and rehabilitating roads to the point where vehicles can easily pass will go far toward promoting market activity. It is estimated that some 80 percent of the road network is in very bad conditions, and most of the rail network is closed apart from a few short-haul services near the ports. The end of civil war has opened the way for a large increase in road transport, and about 50 percent of the primary road network has been reopened to traffic according to government estimates; considering the relative weight of the transport on the final value of farm products and food, the rehabilitation of roads will provide a major benefit to both the agricultural sector and to the consumers, notably those who live in the interior areas of the country and cannot benefit from the cheaper prices of the imported food close to the ports.
- *Strengthening of farmer associations:* Producer associations have proven effective in the Angolan context in supporting marketing by smallholders as well as providing a way to mediate support efforts in terms of credit, extension, and input purchase.
- *Incentives for irrigation:* Despite the dominance of rainfed agriculture in Angola,[61] irrigation can be important to maintain year-round production of food and vegetables. The sharp and rapid transition between the coastal plain and the plateau provides a great number of potential sites to establish reservoirs and major diversion structures to regulate flows and to irrigate the extensive lower floodplains. The morphologies of the central highlands and eastern zone present a high potential for small river diversion structures and small storage tanks. But care needs to exercised in promoting irrigation to avoid uncontrolled and unbalanced expenditures that will not reach smallholders.
- *Address land issues:* The institutions responsible for land tenure, cadastre and registration to ensure greater transparency and protection for the rights of smallholders should be reformed. Special attention should be given to the finalization of the legal framework surrounding land titling based on a participatory

---

61. Three main types of irrigation prevail in the country: (i) large to medium scale irrigation schemes fully or partly equipped with water control works; (ii) small scale gravity or pumped schemes; and (iii) low lands and depressions utilizing water conservation farming practices. According to the available data, there were around 100,000 ha of land that are irrigated in large/medium scheme, of which roughly 39,750 ha are reported to be "operational" in 2004 and less than 20,000 ha for small scale model (against more than 300,000 ha in the pre-war period).

process. Current donor efforts are piloting approaches in land reform, which need to be evaluated for potential duplication and support in additional areas of the country.

- *Decentralized Extension Services:* The Government has recently announced a general plan to decentralize government functions; this deserves a further assessment and financial support at the decentralized level as well as assistance in defining the relationship between extension, farmers and the central government.
- *Agricultural Research:* The principal agricultural research institutions, IIA and IVA, were almost completely destroyed by the war. Reconstituting these as institutions directly responsible to their client population of small farmers is a key element in promoting improved farm technologies; technical cooperation between Minader and other key players in the research of tropical crops, such as IARS, CGIARs and/or the Brazilian Embrapa would also make it possible important improvements and quick gains to the system.
- *Sustainable Management of Natural Resources:* Angola has a tremendous natural resource base in land and water, biodiversity, forestry and fisheries which are critical assets to support sustainable livelihoods and poverty reduction in the country. However, there is a severe limitation of data and information which currently relies on occasional technical studies, selected and deficient government statistics, mostly financed by donors, and thus not necessarily systematized and incorporated in government planning. As the majority of the population in Angola is alienated from the oil and diamonds economy, the sustainable use of natural resources basically means the survival of present and future generations. It is therefore extremely urgent the generation of baseline information about the country's natural resources so that they can be actually managed.

The next chapter discusses palliative measures to improve the welfare of the poor. The recommendations of this Country Economic Memorandum so far have focused on: (i) how to manage appropriately the mineral wealth; (ii) how to improve the business environment to facilitate the diversification of the economy outside of the oil sector; and (iii) how to unleash the potential of the agricultural sector. Addressing all of these issues would bring the Angolan economy close to a path of sustainable and equitable growth. But this will not happen with the speed and the amplitude that many would have hoped, as structural reforms take time to yield concrete visible results. The next chapter discusses instances in which the Government can use palliative measures to improve the welfare and the livelihoods of the poor while the economy moves to a path of sustainable development.

Annex 1. Customs Tariffs and Consumer Taxes Applicable to Imported Agricultural and Related Goods

Item	Merchandise	Current Rates (%)		New Rates[1] (%)		Consumer Tax (%)
		Import	Export	Import	Export	
	Live animals (cow, pork, goats)	5	1	2	1	0
	–pure race reproducers	2	1	0	0	0
	Milk and derivatives (butter & cheese)	5	1	5	1	5
	Roots (potatoes, onions, garlic), beans	5	1	2	1	10
	Fruits (fresh and dry)	5	1	10	1	10
	–banana	10	1	15	1	
	Coffee (green or roasted)	35	1	30	1	30
	–tea	5	1	5	1	5
	Cereals (wheat, maize, rice, sorghum)	2	1	2	1	2
	Flour (maize, wheat, rice)	10	1	15	1	10
	Seeds (oilseeds, grains)	5	1	2	1	2 bulk 5 refin 10 non-ref
	Crude animal & vegetal oil (pork, soya olive, sunflower)	5	1	2	1	2
	–refined in bulk	5	1	5	1	5
	–refined and bottled	5	1	10	1	10
	Canned meat (beef, pork)	20	2	20	2	15
	–fish and seafood	20	1	20	1	15
	Refined sugar (cane & other)	5	1	5	1	10
	–non-refined	5	1	2	1	2
	Canned fruits and vegetables	10	1	10	1	10
	Extracts, essences, spices, sauces	30	2	5	1	0 raw mat. 10
	–yeast	5	1	5	1	5
	Drinks (water, ethyl alcohol)	30	2	30	2	20
	–alcoholic drinks	35	2	30	2	30
	Fertilizers	2	2	2	2	2
	Soaps	5	1	5	1	10
	Farming implements (machetes, scissors)	5	1	2	1	2
	Equipment and machinery –water pumps					
	–coldrooms	2	1	2	1	2
	–scales	10	1	10	1	2
	–earthworks	2	1	2	1	2
	–plows, grades, discs,	2	1	2	1	2
	cultivators, sprayers	2	1	2	1	2
	–tractors	2	1	2	1	2

Merchandises imported are, in addition, still subject to the following charges: ad-valorem stamp tax (0.05%); custom service duty (5%); port charges (EP14 & EP17) estimated at 3%; transport charges (10%).
[1] The new tariff rates were approved as of March 1, 2004. Consumer tax rates remain unchanged.

Annex II. Angola—Scorecard for Rural Agriculture and Environment, 2005

	Score	Comments
I. Macro Environment		
1. Inflation	2	Following the introduction of strong stabilization measures in September 2003, inflation fell to 77 per cent by the end-2003 and 31 per cent by end-2004. In previous years, inflation rates were persistently in excess of 100%.[a,f]
2. Trade Policy	4	The mean import tariff is of 19% with tariffs ranging from 0 to 35% comprised by 9 rates. Certain exports are taxed at rates ranging between 1 to 10%, with an average rate of nearly 4%; crude oil and coffee are exempt. In the case of agricultural products, import tariffs, although generally at or below 5 percent, still apply to many key sector inputs which are not currently produced in Angola, increasing the cost of agrochemicals, machinery and other items. Current tariff structures also offer little protection to the agricultural sector, with rates at 2 percent for many imported food items.[a,b,c]
3. Agricultural and Rural Export Potential	3	Angola has a long history of pre-independence agricultural exports, having once been a major coffee exporter. The country was once the world's third largest exporter on this commodity. Maize was also a major export in the 1970's, amounting to more than 400,000 MT in its peak year, almost all of which was produced by smallholders. However, the civil war resulted in a virtual collapse of marketed production as large numbers of rural inhabitants either fled or reverted to subsistence production. Infrastructure suffered greatly with widespread destruction of roads, bridges and warehouses together with the presence of thousands of land mines in rural areas. Considering the long tradition of the country as a major agriculture exporter in the region, the enormous availability of land and the regularity of rainfalls, the country has a high potential to become again an exporter, as infrastructure rehabilitation moves forward and provided the exchange rate is adjusted in the future.[c]
4. Exchange Rate	1	The current exchange rate, over-valuation of the order of 40% compared to the level of January 1999, has reduced the competitive power of export goods on international markets by making them more expensive to foreigner consumers, and reducing the cost of food imports. This over-valuation thus erodes the export potential of agro-livestock-forestry products, although it has to be admitted that this exchange overvaluation favors imports of inputs essential for the country's post-war reconstruction and recovery. However, it also penalizes labor-intensive kind of agriculture as compared to other countries, such as coffee.[b,c,d]
5. Fiscal Deficit	4	In 2003, the fiscal deficit remained relatively high, at 7.9% of GDP. For the first time, however, fiscal operations included most off-budget operations, which amounted to some 1% of GDP. The fiscal deficit was financed by substantial recourse to external loans and grants and the use of signature oil bonuses. In 2004, the fiscal deficit was reduced to 3.5% of GDP, as a result of oil revenues and measures to improve budget execution procedures and controls.[f]

(*Continued*)

Annex II. Angola—Scorecard for Rural Agriculture and Environment, 2005 (*Continued*)

	Score	Comments
6. Agricultural Expenditures	1	Allocations to Ministry of Agriculture and Rural Development (MINADER) for expenditure in 2004 were Kz2.3 billion (approximately US$25 million). This represents only 0.64 percent of the national total budget, or less than a tenth of NEPAD recommendation amount, i.e., 10%.
		Within MINADER, the bulk of current and programmed investment comprises the rehabilitation of large scale Government irrigation schemes, for eventual hand-over to those occupying the land within the command area. The high proportion of capital expenditure allocated to these schemes contrasts with the declaration of many staff from MINADER concerning the importance of small-scale locally managed irrigation and water control. Moreover, it is not apparent that even if successful in restoring water supplies, such investment will be either economically viable, or will be of benefit to more than a small number of farm households.
		Limited investment has also occurred or been programmed for rehabilitation of Government facilities (research stations, offices). Although no figures were available, it was stated by MINADER personnel that a significant, but unquantified, proportion of the budget has not been authorized over the last years, thus reducing expenditures below the budget figures. As a result, almost all the disbursed budget is destined to salaries, penalizing current costs, such as fuel for the extension workers to visit farmers.[b,c]
7. Corruption	2	Accountability mechanisms exist but are not effective. A High Authority Against Corruption was created by law but has yet to be institutionalized. Lack of transparency and allegations of corruption usually refer to the following areas: (i) the handling and disposition of oil and diamond concessions; (ii) external borrowing practices; (iii) revenue flows in the oil sector; and (iv) procurement practices in the public sector.[a]
8. Accountability and Transparency	2	Progress in this area includes the conclusion in mid-2004 of the auditing of the Central Bank accounts and of Sonangol's financial operations by reputable and independent international auditing companies. In a July 2004 Article IV mission, the IMF noted that recent improvements in transparency, particularly regarding oil revenues, external debt and transactions involving the national oil company, have resulted in some commendable clarification of the government's overall fiscal position and its impact on the economy.[a]
II. Structural Issues		
9. Transport and Power	1	The war and abandonment have taken their toll on the road and rail networks. It is estimated that some 80 percent of the road network is in very bad conditions, and most of the rail network is closed apart from a few short-haul services near the ports. The end of civil war has opened the way for a large increase in road transport, and about 50 percent of the primary road network has been reopened to traffic according to government estimates.

		Since 1975, air transport has provided the only reliable mode of transport within Angola. However, flight safety is substandard, and sometimes precarious, because of the bad condition of many runways, the poor condition or absence of navigation and communications equipment at many airports, and inadequate aircraft maintenance.
		The electricity supply system in Angola has deteriorated during the last 30 years. The main hydroelectric plants and transmission lines of the system were built during the 1970s, and important parts of the system are out of service due to war damage and lack of investment and maintenance. Some areas previously served are today without electricity services. In the areas that still have electricity, including Luanda, power failures and load shedding are frequent, and firms and households that can afford them, rely heavily on small-scale diesel generators.[e]
10. Communications	1–2	In rural areas telecommunications is widely unavailable. While a move towards the liberalization of telecommunications has permitted a vast increase in cell phone teledensity, with better service and lower prices, additional progress is needed in introducing competition by removing barriers to entry. Prices remain higher than for other operators in neighboring countries.
11. Irrigation	1	Despite the dominance of rainfed agriculture in Angola, irrigation can be important to maintain year-round production of food and vegetables. The sharp and rapid transition between the coastal plain and the plateau provides a great number of potential sites to establish reservoirs and major diversion structures to regulate flows and to irrigate the extensive lower floodplains. The morphologies of the planalto and eastern zone present a high potential for small river diversion structures and small storage tanks.
		Three main types of irrigation prevail in the country: (i) large to medium scale irrigation schemes fully or partly equipped with water control works; (ii) small scale gravity or pumped schemes; and (iii) low lands and depressions utilizing water conservation farming practices.
		According to the available data, there were around 100,000 ha of land that are irrigated in large/medium scheme, of which roughly 39,750 ha are reported to be "operational" in 2004 and less than 20,000 ha for small scale model (against more than 3000,000 ha in the pre-war period).
12. Research and Extension	1	Regarding Research & Extension, there is a very limited coverage and low quality services at this point. Links between both services are still very weak.
		The Research Institute is comprised by a network of 12 experimental stations and the system is most severely by shortages of financial and human resources and lack of infrastructure, most of it destroyed by the war. There are basically two centers (in Huambo and Malanje) being rehabilitated and poorly operating, through the support of donors and private sector, including the Bank (EMRP), limited and with serious constraints to disseminate technologies, via the formal mechanisms.
		Currently, there are 72 extension offices in the country, but with all kinds of constraints, from the lack of trained and motivated personnel in the field, precarious conditions of offices, until the inexistence of means of transport to the technicians. The Government has recently announced a very large program to reactive the extension services, through a Chinese financing, totaling some US$ 50 million.[c,d]

(Continued)

Annex II. Angola—Scorecard for Rural Agriculture and Environment, 2005 (*Continued*)

	Score	Comments
II. Structural Issues		
13. Inputs Supply	1	Inputs supply for agriculture is very weak and is left almost entirely to the government, who is expected to import around 20,000 tons of fertilizer this year, and a large share of the seed external purchases. There are some few initiatives from the private sector, but farmers can rarely purchase inputs locally and, even if the supply exists, prices are normally non-affordable due to the high prices resulting for the lack of infrastructure, and the decapitalization of the rural areas. These interventions also usually discourage higher investments on the sector.
		In terms of seeds, some estimates show that Angola produces only 30 percent of its annual demand for seeds and planting materials, and imports are able to fill only an additional 20 percent of demand. Most of the seeds used are produced as grains and normally with low quality.[d,e]
14. Private Sector Environment	1	Angola has for long been characterized as a costly place to do business. This can be attributed to the effects of the long civil war, and that damaged significantly the country's infrastructure, acerbated dependency on natural resources and the associated Dutch-disease-like effects, and the transaction costs in general.
		In terms of the agricultural sector, in particular, there still are some areas where remaining controls on the economy hinder progress in rural development, such as the following: the government is still in the business of setting wholesale and retail margins, which regulations are stated as limits of 25% on each of three market levels, namely producer or importer to wholesaler, wholesaler to retailer, and retailer to consumer; another constraint to the development of private sector in the input market is the continued sporadic distributions of subsidized fertilizer by the government. This unfair and erratic competition dampens long term investments by the private sector.[a,b,c,d]
15. Local Government	2	Although, government has recently announced a general plan to decentralize its functions; this deserves financial support at the decentralized level as well as assistance in defining the relationship between local and central levels. Some key staff in the agricultural sector, such as the Provincial Director of Agriculture and other positions is already nominated by, and respond to, the local Government. However, in the case of extension services, for instance, local staff answers directly to central government, with no accountability to the local administrations in their provinces. Agriculture is perhaps the prototypical example of an activity that is not best managed from a capital city.
16. Community Organizations	2	Despite criticism and successive failures, the spirit of the association in Angola is strong and relevant, and forms part of the lives of subsistence farmers. According to the National Union of Agricultural and Rural Associations and Cooperatives of Angola (UNACA), there are currently around 4,000 peasants' associations and more than 100 cooperatives in the country, mostly concentrated in the central highlands, though present throughout the country. These associations, however, are often referred as very inefficient ones.
		There has been a growing interest by the donor community and NGO to promote a better coverage and quality of local organizations, with some scattered success.[d]

III. Natural Resource Endowment		
17. Land/Labor Ratio	5	Angola has abundant essential natural resources to develop its agriculture, with around 57.4 million hectares agricultural area, of which only around 5-8 million of hectares of arable land are being used. Considering a population in the rural area of around 9 million people, this would provide a very high land/labor ratio.
18. Land Inequality	4	In Angola, all land is state-owned and land use rights can be leased. Disputes over land use are mainly about access to the more fertile and climatically more favored land. However, there is anecdotal evidence that traditional systems of land management are far from being democratic and that the process of allocating leasehold rights to smallholders and or commercial farmers is not necessary transparent.
19. Water Resources	5	Angola is rich in water resources; the amount of surface water in the country is estimated at almost 4,600 m3/sec, or almost 12,000 m3 per inhabitant per year. Most of the available water streams come from upland regions, which could make irrigated agriculture a viable proposition without no need of investments in pumps and electricity, therefore allowing irrigation by gravity. Angola has 47 hydro graphic basins, with most rivers rising in the central part of the Plateau. Twenty-six permanent rivers flow into the sea and others, less long, are important for subsistence farming in coastal communities. Several other rivers, mostly in the south-west, are intermittent, flowing only in the rainy season.[d]
20. Forest Resources	4–5	Of the total 23 million hectares of natural forest the productive area is calculated at 2.4 million hectares. Studies of potential timber production of tropical wood in Angola, on the basis of sustainable management of native forests, show an annual potential production of at least 326,000 m3 in logs, which can be sold both domestically and abroad. Angolan tropical wood has considerable promise for use in panels and wood laminates.[d]
21. Environmental Regulatory Institutions	2–3	The type and dimension of environmental problems grow as the country emerges from the war and recovers. There is no systematic analysis of the causes and effects of environmental degradation, although it has not yet reached alarming proportions, being confined to localities and areas with high population concentrations. However, the absence of monitoring and institutional capacity limits the Government's ability to identify and prioritize these problems.[d]

(*Continued*)

Annex II. Angola—Scorecard for Rural Agriculture and Environment, 2005 (*Continued*)

	Score	Comments
IV. Vulnerability and Risk		
22. Management of Food and Weather related shocks	1	The current situation on the ground suggests that a significant number of recently resettled people in rural areas will remain highly vulnerable, either because they have not benefited from any assistance or because they live in remote and inaccessible areas. Humanitarities agencies, that have provided the larger proportion of food, agricultural inputs and tools to hundreds of thousands of families, are working on exit strategies. Most of them are phasing out their activities on humanitarian aid and moving towards a transition, reconstruction and recovery period. The lack of infrastructure and land mines are also an important limitation to the support of vulnerable people.

Sources used for rating: [a] 2004 WB CPIA Benchmarking: Country Worksheet–Angola; [b] WB Draft Country Economic Memorandum, [c] 2005 WB Draft Agricultural Development Strategy for Angola, [d] 2004 FAO Agricultural Sector Review, [e] 2005 WB MOP Angolan Emergency Multisector Recovery Program, [f] 2005 OECD African Economic Outlook.

Explanatory Notes:

Ratings scale: 1 (low) through 5 (high).
1–Unsatisfactory
2–Moderately Unsatisfactory
3–Moderately Satisfactory
4–Good
5–Excellent

CHAPTER 7

Supporting Livelihoods and Improving Service Delivery

As a post-conflict country, Angola faces a huge challenge to improve the welfare of its population, including the poorest. Angola stands a very good chance of meeting this challenge owing to the increasing resources available from the exploitation of natural resources. In order to succeed on that front, the Government needs to pursue three complementary objectives: (a) support livelihood options and strategies available to people; (b) increase the access of the poor to better quality of social and economic services; and (c) use fiscal savings to improve the quality of public service delivery to the poor. This strategy would support an improved living standard by raising incomes and assets of the poor through the provision of capital, technology, and economic services. It would support the equitable development and maintenance of social and human capital through a community based approach to the provision of social services. This chapter discusses options available to meet these objectives..

Supporting Livelihood Strategies

Promoting sustainable pro-poor growth in a resource-based economy is not easy, but is possible. The record of low income mineral-exporting countries in achieving pro-poor growth is fairly dismal, especially in Africa. Nonetheless, the example of Indonesia shows that it can be done. Key elements in Indonesia's success, are embedded in the recommendations offered elsewhere in this report, including managing inflow of foreign currency to ensure a competitive domestic sector (avoiding "Dutch disease"), supporting rural infrastructure and the rural non-farm sector by using oil-revenues to finance labor-intensive infrastructure projects in rural areas, supporting a better business environment for urban micro-enterprises and the provision of infrastructure and social services to the poor areas as well as the upper income or industrial areas.[62]

---

62. See Timer (2005) for an elaboration of these points.

Different livelihood strategies have been used by the poor to survive in difficult times. People's livelihoods depend on the opportunities and assets (natural, physical, financial, human, and social) available to them. The war not only destroyed the assets of people, ruined infrastructure, and harmed and displaced millions of people. It also significantly impacted the livelihood options and strategies available to people. It ruined the agricultural base of the country, by limiting access to inputs, reducing available labor, and by entirely isolating certain parts of rural Angola. This section of the report discusses different livelihood strategies that have been used by Angolans to survive during difficult times and suggests ways to strengthen the options that have worked well so far.

Oil, Diamonds, and Livelihoods

The oil economy is an enclave with limited impact on livelihood strategies, but the same is not true of the diamond economy. While Angola has recently experienced rapid growth, this has been dominated by the oil sector which has very few forward or backward linkages to the rest of the economy, having therefore few impacts on the livelihood strategies and opportunities available to ordinary Angolans. On the other hand, the diamond sector since the 1990s—when UNITA began its occupation of the mines—has been highly labor-intensive, and has played an increasingly important role for the livelihoods of many informal, artisanal miners, the *garimpeiros*. A majority of Angolans, however, rely on livelihood strategies that fall outside of the mining sectors.

Subsistence Farming

Due to lack of opportunities for most Angolans in the mineral sectors, more than two-thirds of the country's workforce find employment in farming, livestock and artisanal fishing.[63] Indeed, while the last four decades have seen what could arguably be called a rural exodus towards urban and peri-urban areas, an estimated 60 percent of Angola's population still resides in rural areas and more than 70 percent depend on agriculture and rural activities as their principal source of income and food. As is also true in urban areas, however, large numbers of people have lost their assets during the conflict and have had few opportunities for replacing them, and therefore lack sufficient food stocks, seeds, tools and livestock. Without assistance in gaining access to such crucial assets they remain unable to resume normal agricultural production, with little chance of feeding themselves. The Provinces of Huambo, Bie, and Huila—where the largest number of returnees, demobilized soldiers and their families are found—are particularly vulnerable to food security, as are the remote Provinces of Kuando Kubango and Moxico.

Subsistence farming remains the main livelihood strategy for a majority of rural Angolans. Both women and men tend to be involved, with women frequently doing more of the

---

63. In terms of fishing, Angola's coast is very favorable, with a lot of potential both for artisanal and industrial fishing. At independence, Angola was the second-ranked fish exporter in Africa, and even in 1985 produced a total of 400,000 tons. Over-fishing, mainly due to illegal and unregulated fishing by large foreign fishing vessels, has resulted in a rapid depletion of stocks, and put the livelihoods of many coastal villages at risk UNDP (2005: 17).

cultivation, helped by children, while men are more involved in the clearing and preparation of land. However, small-scale farming faces significant difficulties in most areas, such as the lack of inputs (which for the most vulnerable includes labor) as well as difficult access to markets. In almost all cases, farmers are only cultivating a small part of their potential land and yields are low. Perhaps one of the most important assets to the rural poor is access to land, which has been put increasingly under pressure since the war as IDPs, demobilized soldiers and other returnees are entering the market for land.

The presence of a significant amount of land mines and other unexploded ordnances (UXOs), however, is a serious threat. Angola is considered to be one of the countries most affected by UXOs. Estimates for the amount of land mines that litter the country are in the millions, although information on the exact extent of land mine contamination in the country remains limited and unclear. This is exacerbated by the numerous parties involved in mine-laying and the lack of credible records. According to Government figures, however, as many as 4,200 areas are reported to contain or suspected to contain land mines, and are spread across all of the 18 provinces in Angola.[64] Based on a survey conducted from 1996 to 1998 (prior to the last military offensive), as much as 35 percent of Angola's 1,254,000 square kilometers is contaminated by land mines.[65]

Non-farm Livelihoods

The difficulties in making a living from peasant agriculture mean that there is a tendency for people in rural areas to also develop other livelihoods and survival strategies. As in urban areas, large numbers of people survive through casual labor, *biscatos*, of various types, including working on other people's fields, transportation of mud (adobe) or other building materials for construction purposes (mainly carried out by children). Many people also depend on access to, and extraction of, other natural resources. The collection of wood for sale, or for transformation into charcoal and subsequent sale, is a common livelihood strategy, especially in areas near towns or larger roads, or where there is some transport available to urban areas. This practice has led to environmental degradation and depletion of woodlands in these areas. Income from sale of wood or charcoal appears to be low, as barriers to entry to this occupation are low and many families include it in their livelihood strategy, leading to abundant supply and low prices. However, most rural people do not have access to a sustainable income base outside of agriculture, although the coping mechanisms provided through casual labor and collection of firewood remain important complementary strategies. There is thus an urgent need to diversity and expand the agricultural and the non-agricultural base of rural households.

---

64. Government of Angola (2004) "Article 7 Report, Form C", submitted to the Secretary-General of the United Nations on 14 September 2004, in compliance with the Convention on the Prohibition of the Use, Stockpiling, Production and Transfer of Anti-Personnel Mines and on their Destruction (the Ottawa Convention). These figures differ from those provided by the National Institute of Demining (INAD) which indicates that Angola has a total of 4,000 minefields.
65. United Nations Mine Action Service (2005) "Angola Country Overview," available at http://www.mineaction.org.

Urban Livelihoods

There are desperate needs in urban areas that need constant attention. While it is important to design an effective and inclusive rural development strategy, in part just to reduce the incentives for urbanization, there are also desperate needs to be filled in the urban areas that cannot be neglected. Approximately 40 percent of Angola's population currently lives in urban areas, of which about 60 percent (or 20–25 percent of the total population) lives in Luanda. According to UN statistics, the overall urban population is likely to grow to 44 percent by 2015 and to 53 percent by 2030.[66] The rapid urbanization can be explained by both push and pull factors: the war, and dismal livelihoods in rural areas pushed people towards urban zones less affected by the conflict, and the perception of better opportunities, peace and—for some, anonymity—were some of the reasons people were pulled towards the cities. Most of the new residents have settled in the peri-urban *musseques* and occupy a significant part of what used to be mostly industrial and farming areas. Living conditions are extremely poor: people live in overcrowded settlements,[67] with no sanitation or solid waste management. Inequalities in access to services are the source of considerable frustration among inhabitants, in particular poor sectors of society.

Children and Youth

Angola is a young country and as such special attention should be devoted to children and the youth. Currently in Angola, about 67 percent of the population is less than 25. A majority of those who migrated or were displaced to urban centers were youth, making it likely that the percentage of people under the age of 25 living in Luanda and other cities is much higher than the national average. Children and youth growing up during the war find themselves at a particular disadvantage, having had no or little access to basic education and health facilities, key protective factors preventing risky behavior and obviously limiting employment opportunities and social mobility. As a result of increasing pressure on livelihoods opportunities, some youth are turning to criminal activities. An estimated 30 percent of children aged 5–14 years are also currently working (UNICEF 2003). Families often try to improve their livelihoods by putting various family members, including children, to work, either inside the home (thereby sometimes allowing women to work outside the home) or outside.

Access to Land and Livelihoods

Equitable and regulated access to land for agriculture, livestock breeding, and settlement is crucial for economic development in Angola. Control over and access to land is affected by displacement and return, land mines, and by existing regulations and institutions. Risks relate to the appropriateness and implementation capacity of the new land law and its regulations,

---

66. Cited in Jenkins et al. (2002: section 2.2).
67. According to the survey conducted by Development Workshop, 55% of households across Luanda reported over 3 people to a room and 30% reported between 2 and 3 people to a room. For Huambo city, 66% of households reported more than 3 people to a room, and 16% between 2 and 3 people to a room (Development Workshop, 2003: 44, 49).

impact of land mines, and different types of local land conflicts by livelihood. Potential for land conflicts and marginalization relate to settlements in urban areas, agriculture and division of territories in densely populated or mined rural areas, and access to water wells for livestock in the south. Without addressing these issues, it is difficult to visualize a sustainable and successful effort to promote poverty reduction through agricultural growth in Angola.

A new Land Bill has become effective, but there are lingering concerns about its impacts. The bill was passed in August 2004, and published in November 2004, becoming effective in February 2005. The new law replaces old legislation from 1992. In essence, the Law: (i) grants the right to private property in urban, but not rural areas, where right is limited to '*superficio*'; (ii) formally recognizes collectively controlled *community land*, regulated through traditional power structures and community level and good for collateral for community loans; (iii) allows expropriation in cases of existence of natural resources or due to inefficient use of the allocated land; and (iv) imposes requirements on the efficient use of land and conditions on capacity to cultivate. There are concerns, however, that the new legislation focused mainly on increasing the power of state officials to manage land and determine who gets land and on what terms (Development Workshop 2005).

Traditional power structures and its implications for land management are a potential source of concern. First, the power granted to traditional authorities—the so-called *sobas*—may undermine the rights of women. While the law grants equal hereditary rights to men and women, local customs vary. In many communities, a deceased husband's property is seen as belonging to his brothers and nephews. Second, the juridical personality of the communal land is not clear. If it is assigned as individual property to the soba, some fear he may be subject to manipulation, or he himself being tempted to manipulate these powers. Third, there are risks of marginalization of other social groups. With the return of refugees, IDPs, ex-combatants and other members of the rebel forces, community compositions are changing. For those returning to their community of origin, disputes over land will normally be resolved by the soba, who represents the community memory. In other cases, often ex-combatants and abductees do not return to their community of origin but to other communities. Some are too far away, others are inhibited by the stigma or trauma the war has inflicted upon them. Community structures are based on a composition of families. Unless they have family members in their new community, they tend to have difficulties being integrated. Fourth, the institutional setup for the formalization of land rights remains to be specified. Fifth, community ownership is not a concept shared by all provinces and even all communities within provinces.

Strengthening Existing Social Programs

The Government's policy instrument in place to improve the living conditions of the poor is the *Estratégia de Combate à Pobreza* (ECP). The ECP was developed by a group of public officers from the main social ministries under the coordination of the Ministry of Planning, with external technical assistance provided by the World Bank and UNDP.[68] The ECP's

---

68. The Ministries involved were the Ministry of Planning, Social Affairs, Women and Family, Agriculture and Rural Development, Health, Education and Culture, Energy and Water, Fisheries and Environment, Public Works and Transport.

specific objective is to consolidate peace and national unity through sustainable improvement of living conditions of the most vulnerable groups. To this effect, the ECP has defined ten priorities. These are (i) Resettlement/Social reinsertion; (ii) De-mining; (iii) Food security; (iv) (HIV/AIDS); (v) Education; (vi) Health; (vii) Rehabilitation of basic infrastructures; (viii) Employment and vocational training; (ix) Governance; (x) Economic management. The ECP budget is $3.2 billion for a period of five years (2003–2006) but does not reflect the considerable increases in resources which are now being made available to the budget. In revising the ECP to take into account the larger fiscal envelope with the recent windfall gains, the authorities should also cast the associated public spending plans within a medium-term strategy for poverty reduction and for achieving the Millennium Development Goals (MDGs).

The Government efforts to tackle the needs of the poor and vulnerable groups are being addressed in national programs. These are in the areas of Social Reintegration (ADRP); HIV/AIDS (HAMSET); Reconstruction and Institutional Support (PAR); Reconstruction/Rehabilitation (FAS). A summary of all of these programs can be found in Box 7.1 and a detailed description of the ECP's priorities as well as an assessment of the performance of the national programs (which is beyond the scope of this report) is presented in the *Country Social Analysis* for Angola, which is a background report to the *Country Economic Memorandum*.

Other social programs that are neither nationwide nor being administrated by Government entities are also relevant. These are being funded by bilateral agencies, United Nations Agencies, and the private sector, and are implemented by international NGOs.

The main program funded by a bilateral agency is the *Luanda Urban Poverty Program (LUPP)*. The LUUP benefits from a grant from the Development Fund for International Development (DFID), and was created in 1998. It is being implemented by a consortium of international NGOs (Development Workshop, Save the Children UK and Care International). The main objective is to contribute to poverty reduction in Luanda peri-urban areas while testing best practices for participatory local governance, through multi-sectoral approaches through education, water and sanitation and good governance. The province of intervention is Luanda and its peri-urban areas (Sambizanda, Kilamba Kiaxi, Ho Yi Ha Henda). The budget for the first phase (1998–2003) and second phase (2003–07) is of £ 9 million respectively.

The private sector is also supporting social programs. These are small scale but do complement other actors' activities in the sector of agriculture and the provision of microcredit. Chevron-Texaco Corporation, a multinational oil company, along with other partners such as USAID, is contributing to the agricultural sector through an integrated development program called Angola Partnership Initiative (API) that aims at: (i) promoting rural development through food relief, re-integration of internally displaced persons, ex-combatants, refugees, food production, seed multiplication, rehabilitation of basic infrastructure, and rural credit; (ii) promoting the development of rural small and medium enterprises through microfinance, vocational training and business development; and (iii) supporting the rehabilitation effort of the education sector. The program is estimated at US$58 million and is being implemented by international and local agencies and NGOs specialized in agriculture and rural development projects, such as CARE, World Vision and ADRA. The provinces of intervention are Huambo, Bie, Kwanza Sul and Huila. The program has duration of four years (2003–07).

Box 7.1: Existing National Programs Addressing PRS Objectives

The Angola Emergency Demobilization and Reintegration Project (ADRP) benefits from a World Bank Trust Fund and is administrated by the Ministry of Social Affairs and the National Institute of Vocational Training for the Reintegration of Ex-Soldiers (IRSEM). The program has for development objective the demobilization of 105.000 ex-military of the UNITA and 33.000 of the Angolan Forces (FAA) and the support of their social and economic reintegration in civilian life; (ii) to facilitate the redistribution of the Government' expenditures of the military sector for the social and economic sectors. To accomplish these ADRP comprises four components: (i) demobilization; (ii) reintegration; (iii) support to special groups; and (iv) support to institutional development and the implementation of the program. The program is implemented by the local IRSEM and national NGOs. The duration of the program is of four years (2003-2006) with a budget of US$45 million. The provinces of intervention are Benguela, Huambo, Hu–la, Bie, Kwanza Sul and Malange.

The HIV/AIDS, Malaria and Tuberculosis Control Project (HAMSET) benefits from a World Bank grant and is administrated by the Ministry of Health. The program has for development objectives (i) reduce the spread of HIV/AIDS in the Angola population through multi-sector approach that strengthens institutional capacity and increases access and utilization of health services for prevention, diagnosis, care and support; (ii) strengthen the capacity of the health sector to detect new cases of tuberculosis, improve treatment compliance, and increase the completion rate; (iii) strengthen the capacity of the Ministry of Health for effective case management of malaria. To this end, the project comprises the following four components: (i) public sector response; (ii) health sector response; (iii) community response; and (iv) project coordination. The program is implemented by the local provincial health delegations and national NGOs. The program has duration of five years (2005–2010) with a total budget of 39.6 million. The provinces of intervention are Luanda, cabinda, Benguela, Cunene, Lunda Sul, Kuando Kubango and Lunda Norte.

Rehabilitation and Reconstruction Program (PAR) benefits from a European Union grant and is administrated by the Ministry of Planning. The primary objective of the priority phase of the program are (i) to consolidate peace; (ii) to strengthen national reconciliation and improve security by extending state administration to all areas of the country; (iii) to achieve rapid improvements in food security and living conditions for the most–vulnerable and war–affected groups, especially those suffering from malnutrition and disease; (iv) to create safe and adequate conditions for the return of displaced people to their regions of origin; and (v) to pursue the reconstruction of social services. There are four implementation components: (i) social sectors and rural development; (ii) priority rehabilitation of critical infrastructures; (iii) capacity building, institutional strengthening and sector development strategies; and (iv) management and monitoring of the program and preparation of the second phase. The areas of intervention are 31 municipalities of the central highlands of Angola, namely in the provinces of Benguela, Huambo, Bié and Huila. The program has already implemented two phases. The Project has a budget of EUROS 52.000.000 and duration of 8 years (1997–2005).

The Social Action Fund (FAS) benefits from a World Bank loan and a trust fund from the European Union and is administrated by the Ministry of Planning. FAS was created on October 28, 1994 by the Executive Decree of the Council of Ministers no. 44/94. It has already benefited from two phases. The third phase development objective is to reach, improve, and expand the social and economic use of basic infrastructures applying a development driven approach that has for objective of building the human and social capital within and among communities with the external agents' support as the local administrations and organizations of the civil society. The main sectors of interventions are : Education; Health; Water; Sanitation; Economic and Productive and environment. FAS III intervenes in 11 provinces of Angola with the intention to cover the national territory until the end of the third phase in 2008. The provinces of intervention to date are Cabinda, Luanda, Benguela, Zaire, Bengo, Kuanza Sul, Huambo, Bié, Huíla, Namibe, and Cunene. FAS III has an estimated budget of US$120.000.000 for period of 2003–08. The FAS has already disbursed over US$77 million.

Source: World Bank (2005), *Country Social Analysis,* Background Report for the Country Economic Memorandum.

There are also a plethora of international NGOs that are active in the delivery of basic social services. A leading project in the area of micro finance is the NOVO BANCO project. Its main objective is to create a sustainable, profitable commercial microfinance institution to provide credit and other financial services to micro, small and medium sized businesses. NOVO BANCO targets enterprises and entrepreneurs in the trade, services and production sectors working in urban and peri-urban communities. The project is being implemented by Frontier International in the province of Luanda with the intention to expand to other provinces. The project has a capital of US$4.9 million and a total fund of US$7.9 million.

Local NGOs have difficulties to implement large-scale development programs given the donor-driven nature of their operations. Moreover, the majority of NGOs have limited experience in sustainable development activities given the fact that they have only been engaged so far in the delivery of humanitarian assistance in an emergency context. Although the scope of their programs is now adapting to the transition to the rehabilitation and reconstruction, the majority have sustainability problems and therefore can only operate with external funds.

Churches also play a very important role in providing services to the poor. Both the Catholic and Protestant Church have the largest number of adherents, given that there are over 80 per cent of Christians in Angola. These Churches are active in responding to the social needs of its membership through a national network of Caritas offices that operates in the provinces, and departments that promote community development and social programs. The four Protestant churches and church-related organisations that seem to most actively promote these types of programs are the *Igreja Kibanguista*, the *Igreja Batista*, the *Associacão Evangelica de Angola* and the *Conselho de Igrejas Cristas*.

All of these programs have had some success, but more needs to be done to reach the most remote and most needy areas of the country. Although existing national programs administrated by the government and the private sector suggest that the intended objectives are being implemented with some degree of success, the majority lack capacity to fully respond to the poor given that the demand far outstrip the supply. However, it is also evident that the limited access to more remote areas of the country due to communication constraints and mines have contributed to the fact that over eight provinces such as Lunda Norte, Lunda Sul, Moxico, Kwando Kubango, Uige, Kwanza Norte, Zaire and Malange have limited or non-existing presence of activities by these programs.

Reaching the Poor with Social Services

One of the surest ways to use oil revenues to improve the welfare of the poorest is to reduce the overall social service deficit. Limited information available suggests that this will require a restructuring of the budget, and possibly some changes in the budget process. In terms of expenditure by sector, Angola still spent 40 percent of its budget on defense and national security in 2004. This limits the amount available for sectors such as health, education and infrastructure (water supply, roads, and so forth). As a result, Angola spent only 14 percent of the budget on education, 13 percent on health and social assistance, and less than 5 percent on roads, water and sanitation, and urban infrastructure. By contrast, in the same year, Uganda (which faces significant security problems in the northern districts, as well as a high debt burden), spent 19 percent on education (down from 21 percent in the previous year), 12 percent on health, 8 percent on roads and works and 3 percent on rural water supply. Uganda achieved this by reducing security (including defense) expenditures to 11 percent of the budget. This expenditure is still high compared with its neighbors, but as noted above, the conflict in the north has taken a large toll, both human and fiscal. Within expenditure categories, Uganda has been able to direct expenditures toward basic services, especially in under served areas. In 2003/4, 60 percent of the education budget goes to primary education, compared to 11 percent of non-wage education expenditures to primary education in Angola in 2004.

To improve equity in the provision of social services, budget allocations should be based on unit costs and more spending autonomy extended to provincial governments. In terms of non-defense expenditure by province, Angola also has wide gaps. As there is no data on, for example, how many children in each province go to school, one cannot calculate average education spending levels per child. In addition, data is only available on expenditures for the portion which is delegated to the province to spend (86 percent of the total). By calculating expenditures per capita for a few provinces one can find quite a gap (see Figure 7.1). The variance in education appears to be 10 times higher than the average. This is quite high given that the province is already a fairly large agregation. In health, the proportion delegated to the provinces is much less (only 48 percent). This may be why the expenditures on health appear more equally distributed (see Figure 7.2). Countries such as Tanzania and Uganda have reduced these regional gaps substantially by allocating more expenditure authority to local governments, and by allocating funds on a per enrolled child basis. This not only ensures a more equitable distribution, but it also encourages teachers and principals to maximize enrollment of children in the area.

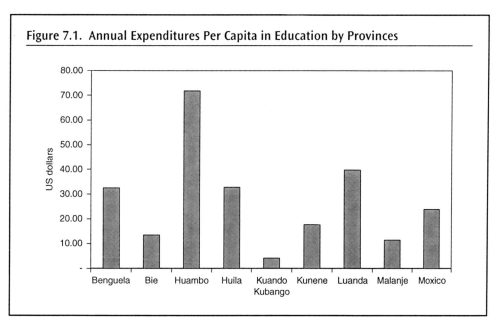

Figure 7.1. Annual Expenditures Per Capita in Education by Provinces

Note: Excludes wages and other central government expenditures.

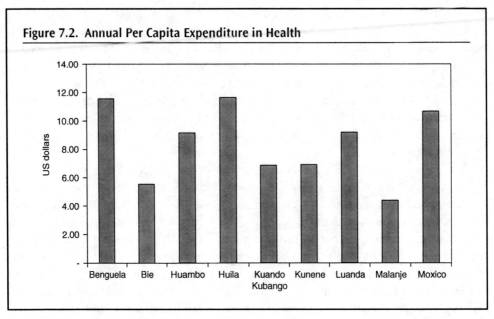

Note: Excludes wages and other central government expenditures.

Mechanisms to consult the poor on how and where to allocate public funds can be very helpful. African countries have also found that increased community participation in the design and implementation of social service investments as well as in operation and maintenance results in more efficient and effective public spending on social infrastructure. Angola's experience with FAS is a very good example of this mode of operation. As part of its ongoing decentralization process, Angola may wish to consider incorporating elements of this delegation of responsibility in its programs. Mechanisms to automatically consult the poor in regards to decision-making on public investment should be put in place in addition to accountability channels in order to ensure quality control and civic engagement. The poor quality of services offered, has led to the lack of trust in the institutions by the poor and the over reliance on the informal sector for social services. However, evidence indicate that this option is not always cheaper and/or of better quality.

Effective social sector spending requires predictable resources available to spending units, especially in the labor-intensive sectors such as health, education, and social assistance. Angola has not yet achieved this goal. While detailed budgeting is done every year, only 72 percent of what was allocated in the budget was made available to spending units in 2004. The discrepancy between budget allocation and what was received by spending units also varies substantially by province. In Uige, Zaire, and Luanda Norte, the amount spent was less than 40 percent of the budget allocation. Other provinces did much better as in Benguela, 68 percent of the budget allocation was spent. As a result, employees such as teachers often face arrears in pay, and are forced to take a second job, sometimes resulting in absences form the classroom. As the economy settles down and revenues become more stable and predictable (especially oil revenues, the main revenue source for the Government), the Government could set targets toward more realistic budgeting. An oil revenues stabilization fund may help in this process.

To increase the effectiveness of current activities, a higher degree of coordination among development partners and the government should be put in place. Activities implemented by national programs are providing many opportunities for consolidating partnerships to improve the livelihoods of the poor and vulnerable. However, their implementation is hampered by the lack of a national coordination mechanism. The majority of national programs do not report systematically on their lessons learnt among themselves and partnerships are therefore taking place mostly at the local level and on a project basis without the umbrella of a national coordinating agency to ensure that the implemented activities respond to the ECP. In the long run, this could create a situation in which the overall impact of the several existing programs happens in an isolated fashion to the detriment of the strengthening of social protection measures and the design of well informed policies. To address this problem, the government should create a coordinating agency and encourage the establishment of a mechanism for information sharing in view of informing policies.

Inequality of opportunities causes certain groups of Angolans to face greater barriers to access basic services than others. Data from the 2002 MICS survey suggest that the ability to gain access to basic social services is directly influenced by the household's income level. Access to primary education, for example, is only 35 percent for the poorest quintile, while for the richest quintile it is more than double that, at 77 percent.[69] Literacy rates compare at 62 percent for the poorest quintile and 95 percent for the richest quintile. Gender disparity is also apparent, with female literacy rates within the poorest quintile of 27 percent compared with 62 percent for men, and 86 and 95 percent within the richest quintile, respectively.[70] Government data from 2002 show that overall completion rates for primary education for girls are estimated at 41.3 percent compared with 56.8 percent of boys.[71]

Another equally important challenge is how to offer basic social services to all at accessible costs. Not everybody in Angola is able to go to a hospital without experiencing tremendous difficulties. In the eastern Provinces, for example, where the humanitarian needs are perhaps the most striking, distances to the nearest hospitals can sometimes be more than 100kms, and roads are not always there to make matters worse.[72] Access to healthcare is also prohibitively expensive. Again, in Hoji ya Henda, treatment at a public hospital can cost from Kz500 to Kz1,500 (roughly US$6 to US$18), many times the daily income for most of Luanda's population.[73] Public school places in Luanda, moreover, tend

---

69. Conditional cash transfers (CCTs) have been used in many countries to induce poor parents to enroll their children in school. CCTs make payments to poor families, typically mothers, on the condition that children attend school regularly. The programs can be seen as compensating for the opportunity cost of schooling for poor families and represent one approach to addressing failures in credit markets and the imperfect agency of parents. The biggest programs of this kind are Oportunidades (formerly PROGRESA) in Mexico, the Bolsa Escola in Brazil, and the Food for Education Program in Bangladesh (see WDR, 2006, p. 137).

70. The 2006 World Development Report argues that gender inequity directly affects the well-being of women and decisions in the home, affecting investments in children and household welfare (p. 51). Econometric evidence cited in the report confirms that an increase in a woman's relative worth and an improvement in her fallback options have positive effects on the children (p. 53).

71. Progress report on MDGs (2005: 24). Gender inequities are also reflected at the government and national political level, where women only account for 11 of 70 ministers and vice-ministers and only 14% of the national assembly. At the Provincial level, there are currently no female Governors or Vice-Governors.

72. Robson (2005).

73. Yngstrom (ibid).

to be limited. A range of private schools or classes are emerging to fill the vacuum, some legal some not, both more expensive than the state school system.

A carefully designed and phased public works programs can potentially help to support agricultural growth and thus contribute to poverty alleviation. To support the rural economy to get back on its feet, the Government may wish to consider complementing an agricultural support package (which should increase productivity and reduce co-variate risk) with an explicitly targeted public works jobs program for the low skilled labor force. Targets for the program would include return migrants, who may have trouble getting land or integrating into the farming sector as they have been away from farming for such a long period of time. Farmers would also want to participate during the off-season. In countries such as Tanzania, Senegal and Malawi, these programs have been implemented through social fund structures, drawing donor support in to help finance the programs. In South Asia, where local government structures are more developed, these programs have been implemented through Government structures with some success. Once the rural economy has stabilized in an area, programs to support non-farm micro and small enterprises can be introduced and the jobs programs phased out.

All things considered, improvements in social indicators should be the main objective of any public policy in Angola. Child mortality is one of the highest in the world as one out of every four children is expected to die before the age of 5. This estimate does not vary much between rural and urban areas. Nutrition indicators are equally poor: 30 percent of children are moderately underweight, and 8 percent were found to be severely underweight. Likewise, 45 percent suffer from moderate stunting (low height per age) and 22 percent suffer from severe stunting, a very high level. In the lowest asset quintile, these numbers are double the ratio for the highest quintile, but even in the highest quintile the numbers are high: 13 percent severe stunting. By comparison, Mozambique in 2003 recorded a slightly lower level of moderate stunting, but a much lower percentage of severe stunting.

Recent Progress in Service Delivery

Despite difficulties with logistics, there has been some limited progress in the areas of education, health, and resettling of displaced people. Reportedly, some 29,000 additional teachers have been recruited, and one million more children are apparently now in school. Polio is fast getting closer to polio-free status and towards its goal of reducing child mortality rates. In terms of access to populations in remote areas, dozens of bridges and nearly 1,000 km of roads have been reconstructed and rehabilitated. And although more than a million Angolans still require food aid, almost half a million more hectares of land have been cultivated since the peace. Logistics have hampered most efforts to improve services in rural areas. The Government, supported by international assistance and NGOs, is resettling displaced persons and ex-combatants back to their home areas, resulting in a migration out of Luanda and other large cities. This may help to improve the subsistence level of the poor compared with living in IDP camps if farmers can earn an adequate subsistence off the land, but it may also increase exclusion and the rural urban-gap owing to the difficulties Angola faces in improving services in poor areas.

Recent efforts to improve service delivery have yet to reach the most remote areas. In 2004, the government budget was approved with significant increases in allocations towards health, education and welfare services (an increase of 9 percent compared to the

Table 7.1. Public Services According to their Perceived Importance
(1 = Maximal; 7 = Minimal; %)

Services	1	2	3	4	5	6	7
Schools	92.29	3.15	1.82	0.77	0.70	0.35	0.91
Electricity	82.84	6.93	3.43	2.59	1.40	0.91	1.89
Roads	68.19	9.39	9.18	3.50	1.40	1.12	7.22
Health Center	86.19	7.01	2.80	2.10	0.56	0.70	0.63
Sewage	72.48	6.65	7.21	3.64	1.75	2.10	6.16
Home Tap Water	83.31	8.63	3.16	1.61	1.05	0.98	1.26
Pub. Transportation	76.30	7.10	2.70	2.34	1.77	2.27	7.52

previous year). Provincial Governments, however, still lack sufficient funds and technical capacity to improve service delivery without external assistance. With limited capacity of the Government to deliver services, the main providers of basic social services, humanitarian relief and development resources have since the early 1990s been non-governmental organizations (NGOs), supported by international donors (Pacheco, 2005). In urban areas some services have also been provided by the private sector (mainly water). Nevertheless, NGOs have so far tended to concentrate in particular areas of the country—and not necessarily the ones with the greatest needs—and are very rarely sufficient to meet local needs. Logistics continue to create difficulties for improving access to services in remote areas. As a result, there is a risk that the low level of services in these areas will continue accentuating the inequalities between areas and between social groups.

Using Fiscal Savings to Improve Service Delivery

Phase Out Subsidies to Obtain Fiscal Savings

Eliminating fuel and utility subsidies can be a viable option to improve service delivery and the welfare of the poor. The government is considering a gradual elimination of fuel prices to be followed by the phasing out of subsidies to public utilities. The World Bank, with the support of the UK Embassy in Angola, commissioned a study to assess the welfare impacts of such policy. The conclusions of this study are very telling. In the first place, the population ranks very highly the types of service provided by the government, but the quality of these services is considered by the majority of the population to be of minimal quality (see Tables 7.1 and 7.2).[74] In the second place, subsidy provision in Angola does not target the

---

74. In the authorities' view, the school system in Angola has been a major priority for the government as it remains highly ineffective and incapable of forming skilled labor in the pace which would be required to satisfy the current demand for good education by both households and emerging businesses.

Table 7.2. Satisfaction Rates with Public Services (1 = Maximal; 7 = Minimal; %)

Services	1	2	3	4	5	6	7
Schools	19.55	10.16	16.82	8.27	4.70	7.57	32.94
Electricity	12.46	6.72	9.10	12.46	6.02	9.87	43.35
Roads	6.74	3.58	4.28	9.89	8.42	11.16	55.93
Health Center	15.16	9.54	14.11	12.42	7.58	10.46	30.74
Sewage	8.63	2.18	5.89	8.91	5.89	7.16	61.33
Home Tap Water	18.79	5.89	6.10	11.43	5.12	9.05	43.62
Pub. Transportation	15.33	6.53	9.01	9.87	6.39	13.20	39.67

poor who are most likely to be excluded from both public service provision and consumption. Therefore, the marginal negative social impacts of removing fuel and utility price subsidies will likely be small.

The phase out of fuel and utility price subsidies represents an extra opportunity to improve spending in poverty reduction and development programs. Subsidy expenditure was the main source of expenditure within the umbrella of governmental transfer payments. Of total transfer payments, subsidy expenditure accounted for 66.57 percent in 2001, 74.09 percent in 2002, 67.31 percent in 2003 and 76.86 percent in 2004. Between 2001 and 2004, subsidy expenditure expanded by 377 percent. As subsidy expenditure amounted to US$1.286 billion in 2004, the savings to be had with the complete elimination of subsidies are tremendous. Clearly, there would be enough resources to compensate poor households for increased burdens associated with higher fuel prices.

Subsidies to fuel and utility prices should be phased out gradually and the savings redirected to improve the quality of public services and to compensate the poor. In a first stage, the authorities will need to announce a comprehensive program to deal with subsidies in a phased strategy beginning in 2006. Some limited increase to show intent would be a good way to start the program, but the announcement of the program would have to state clearly that the initial objective is, for example, to align fuel prices with the 2006 budget's implicit oil price (US$45/barrel). A larger increase could be introduced in mid-2006 matched with a program to spend the funds saved through a compensation program and on needed infrastructure at the same overall fiscal deficit. In a second stage, the current pricing mechanism used by Sonangol should be reassessed and the adjusted level of subsidies phased out gradually while the compensation program designed in the first stage continues to be implemented. At any point in time, the program should be adjusted in case there is a crash in international oil prices.

How to Manage Political Costs

In contemplating the implementation of a policy of gradually and periodically adjusting fuel and utility prices, Angolan authorities should not neglect the policy's political costs. This is particularly true now in the eave of major elections. International evidence demonstrates that fuel price hikes unaccompanied of a set of palliative measures typically leads to public protests and may trigger violence and social unrest, even at times following elections. The incentives of political contestants of generating public debate and criticism of fuel and utility price increases at times preceding elections are heightened. Incumbents anticipate this and thus know that if they support fuel price increases, they may end up losing high numbers of votes to their competitors. The experience of other countries demonstrates that any program to reallocate fiscal spending should pay close attention to the following broad principles:

- *Political attractiveness:* Off-setting, politically attractive expenditure programs are key to the political and social sustainability of a subsidy removal. A high visibility and credible announcement of compensating measures is an indispensable feature of the package.
- *Pro-poor and effective targeting:* Measures should ideally be those that maximize development and poverty reduction impact. "Compensation" programs should be targeted to benefit the poor and possibly other key losers of the subsidy removal as well. They need to be seen to do so as well.
- *Speed of spending and impact on households:* The speed at which programs are designed and money is spent is important for macroeconomic reasons, while the speed at which programs impact poor households is important for compensatory and political economy reasons.

Palliative Measures, Monitoring, and Evaluation

One of the palliative measures that may bring large benefits to the poor is a water and sanitation policy intended to provide access to higher quality water and sanitation services. The welfare analysis carried out in the World Bank-UK-sponsored study on the impacts of phasing out subsidies indicates that changes in water price will bring about a much larger marginal social impact than changes in fuel prices. Given the precarious state of affairs in water infrastructure, management and service delivery, there appears to be ample room for public policymakers to reform the public water sector with the goals of making it more efficient, expanding water supply to the population and improving water quality. Such a reform, if successful, will naturally lead to a substantial reduction in the water price faced by the average consumer in Luanda. The funds needed to finance such a large venture could come from funds saved with the gradual phase out of fuel subsidies and from private sector participation.

Other palliative policies that can be implemented are as follows:

- *Public Education Policy:* Promotion of voucher systems, conditional cash transfers and school meals in order to offer incentives to parents to send their children to school and incentives for children to stay in school;
- *Health Care Provision Policy:* Promote infrastructure investment, offer preventive care and mobile clinics to expand health care service access;
- *Public Transportation Policy:* Creation of a social pass which can be used in buses and vans interchangeably, cost subsidization of van service based on utilization rates, implement employer-sponsored passes.

A successful package to phase out subsidies in Angola will demand political commitment and a good implementation and monitoring strategy. In order to successfully implement the package the GoA will need to form teams and attribute responsibilities. The teams should cover the following main areas: (i) macroeconomic and fiscal issues; (ii) expenditure programming (social protection and compensatory expenditure, and other development and poverty reducing expenditures); and (iii) socialization of savings. They would have to work together and consult each other in order to develop initial proposals based on their own experience and on the findings of the Bank-UK sponsored subsidies study.

Finally, in order to know if a poverty reduction strategy is effective in reducing poverty, it is necessary to set in place a poverty monitoring system to track key indicators over time and space. In Angola, there is still a dearth of knowledge on the livelihoods of the poor. Overall, it is difficult to have a full assessment of the existing safety nets programs given that these do no have at the outset an expected number of beneficiaries. This failure comes from the lack of reliable data on poverty to which programs could refer to in order to assess appropriateness and effectiveness in their targeting and the impact of their activities. A sensible way of tackling this limitation would be to strengthen the National Statistical Institute (INE) and establish a program of annual monitoring of poverty and indicators and disseminate these to the public at large. The World Bank stands ready to assist the GoA in each of these areas.

Statistical Appendix

Angola: Country Data

Area (Thousand Sq. Km.)	Population	Density
1,246.7	14.5 million (mid 2005) Growth rate: 3.0% (average 1998–2005)	11.6 per square km.

Millenium Development Goals: Selected Indicators

	Latest year available between 1998–2005	Sub-Sahara Africa
GNI per capita (Atlas method)	1,350	600
Net primary school enrollment (% age group)	61	64
Ratio of girls to boys in education	84.1	83
Under 5 mortality rate (per 1,000 live births)	260	168
Life expectancy at birth (years)	47	47
Access to improve water sources	38	58
HIV Prevalence rate, female (% age 15–24)	3.9	9.4

Gross Domestic Product in 2005

	US$Million	%
GDPmp	32,810	100.0
Investment	2,473	7.5
Gross domestic savings	10,759	32.8
Resource balance	8,286	25.3
Exports of goods and services	24,120	73.5
Imports of goods and services	15,834	48.3

Government Finances as percent of GDP (%)

	2002–04	2005
Current revenue	37.5	38.0
Current expenditure	33.2	25.2
Current balance	4.3	12.8
Capital and other expenditures	8.6	6.0
Overall balance (accrual basis)	−4.3	6.8

Annual rate of growth of GDP at constant 1997 prices (%)

2000–2001	2002–2003	2004–2005
3.1	8.9	15.9

Money and Credit
In billions of Kwanza

	2001	2002	2003	2004	2005
Net foreign assets	18.0	46.0	60.1	123.8	185.3
Domestic credit	12.3	28.0	38.2	10.0	−14.3
Money and Quasi-money	25.5	65.6	70.9	66.7	146.6
Other net liabilities					
M2/GDP (%)	21.0	21.5	17.1	14.8	13.7

Inflation and exchange rate

	2001	2002	2003	2004	2005
Period average inflation (%)	152.6	108.9	98.3	43.6	23.0
Kz per US$	22.1	43.7	74.6	83.4	87.2
US$ per Kz	0.0453	0.0229	0.0134	0.0120	0.0115

Balance of payments
Million US$

	2001	2002	2003	2004	2005
Exports, f.o.b.	6645	8166	9515	13474	23724
Imports, f.o.b.	3179	3760	5480	5832	8667
Services (net)	−3316	−3115	−3120	−4480	−6770
Income (net)	−1561	−1635	−1726	−2484	−4075
Current transfers (net)	91	32	99	7	30
Financial and capital account					
Capital transfers (net)	4	10	0	11	6
Direct investments (net)	2146	1643	1652	1414	1639
Medium- and long-term loans	−618	−162	178	906	1408
Other capital (net, incl. errors and omissions)	−985	−1742	−1005	−1970	−5512
Overall balance					
Net international reserves (−increase)	440	229	−443	−1233	−2117
Exceptional financing	334	334	330	187	335

Merchandise exports

	2002–04	%	2005	%
Total	10385.0	100.0	23723.7	100.0
Crude oil	9453.1	39.8	22307.7	94.0
Refined oil products and gas	145.9	0.6	245.0	1.0
Diamonds	738.8	3.1	1089.2	4.6
Other	47.2	0.2	81.9	0.3

External Debt in billion of US$ (end period)

	2005
Total outstanding and disbursed	12.6

Debt service to net export ratio (%)

	2005
	10.5

Angola—Key Economic Indicators

Indicator	Actual			Estimated			Projected	
	1999	2000	2001	2002	2003	2004	2005	2006
National accounts (as % of GDP)								
Gross domestic product[a]	100	100	100	100	100	100	100	100
Agriculture	6	6	8	8	8	9	7	9
Industry	73	72	65	68	67	66	74	72
Services	21	22	27	24	24	25	19	19
Total Consumption	79	58	85	76	81	75	67	65
Gross domestic fixed investment	29	15	13	13	13	9	8	14
Government investment	13	6	6	7	8	5	5	12
Private investment	16	9	7	6	5	4	3	2
Exports (GNFS)[b]	86	90	77	74	70	70	74	70
Imports (GNFS)	93	63	75	62	63	54	48	49
Gross domestic savings	21	42	15	24	19	25	33	35
Gross national savings[c]	−1	24	−1	10	8	13	20	23
Memorandum items								
Gross domestic product (US$ million at current prices)	6153	9135	8936	11386	13956	19800	32810	44103
GNP per capita (US$, Atlas method)	450	480	540	720	770	960	1350	1980
Real annual growth rates (%, calculated from 1997 prices)								
Gross domestic product at market prices	3.2	3.0	3.1	14.5	3.3	11.2	20.6	14.6
Gross Domestic Income	21.2	48.0	−7.7	14.4	6.6	21.3	52.0	26.5
Real annual per capita growth rates (%, calculated from 1997 prices)								
Gross domestic product at market prices	0.6	0.2	0.2	11.2	0.2	7.7	16.5	11.6
Total consumption	2.1	9.6	17.5	−3.8	5.0	7.0	25.7	−13.6
Balance of Payments (US$ millions)								
Exports (GNFS)[b]	5310	8187	6848	8373	9716	13797	24120	30785
Merchandise FOB	5157	7920	6645	8166	9515	13474	23724	30335
Imports (GNFS)[b]	5704	5739	6697	7082	8801	10635	15834	21640
Merchandise FOB	3109	3040	3179	3760	5480	5832	8667	12514
Resource balance	−394	2448	150	1291	915	3162	8286	9145
Net current transfers	56	28	91	32	99	7	30	28
Current account balance	−1710	795	−1320	−312	−713	685	4242	3896
Net private foreign direct investment	2472	879	2146	1643	1652	1414	1639	2237
Long-term loans (net)	−291	−766	−618	−162	178	906	1408	2397
Other capital (net, incl. errors & ommissions)	59	−276	−647	−1398	−675	−1772	−5172	−3419
Change in reserves[d]	−530	−631	440	229	−443	−1233	−2117	−5112

(Continued)

Angola—Key Economic Indicators (*Continued*)

Indicator	Actual			Estimated			Projected	
	1999	2000	2001	2002	2003	2004	2005	2006
Memorandum items								
Resource balance (% of GDP)	−6.4	26.8	1.7	11.3	6.6	16.0	25.3	20.7
Real annual growth rates (YR97 prices)								
Merchandise exports (FOB)	1.4	2.0	−0.9	21.3	−2.7	14.1	28.5	13.2
Merchandise imports (CIF)	55.3	4.8	7.9	15.4	29.2	−3.7	43.5	45.5
Public finance (as % of GDP at market prices)[e]								
Current revenues and grants	46.4	50.2	45.1	38.3	37.9	36.9	38.0	38.0
Current expenditures	50.6	43.5	35.6	35.0	36.3	30.6	25.2	23.4
Current account surplus (+) or deficit (−)	−4.2	6.7	9.5	3.3	1.6	6.3	12.8	14.5
Capital and other expenditure	31.1	15.1	13.1	12.3	8.0	7.9	6.0	12.3
Total expenditures	81.7	58.6	48.7	47.3	44.3	38.5	31.2	35.7
Overall Balance (accrual basis)	−35.4	−8.4	−3.6	−8.9	−6.4	−1.6	6.8	2.2
Monetary indicators								
M2/GDP	22.8	17.3	21.0	21.5	17.1	14.8	13.7	15.8
Growth of M2 (%)	680.9	303.7	161.2	158.5	66.3	37.5	60.0	43.2
Price indices (YR97 = 100)								
Merchandise export price index	99.0	149.2	126.2	127.9	153.2	190.0	260.3	294.2
Merchandise import price index	92.7	86.5	83.8	86.0	97.0	107.1	111.0	110.2
Merchandise terms of trade index	106.8	172.4	150.6	148.8	157.9	177.4	234.6	267.1
Real effective exchange rate (US$/LCU)[f]	71.0	85.3	96.4	98.1	116.1	139.1	173.3	—
Consumer price index (% change)	248.2	325.0	152.6	108.9	98.3	43.6	23.0	12.9
GDP deflator (% change)	556.9	418.2	108.5	120.5	102.5	42.7	43.5	8.0
External debt								
External debt in billions of US dollars	10.3	9.4	9.2	9.2	9.7	10.6	12.6	14.8
External debt to GDP ratio	167.4	102.9	103.0	80.8	69.5	53.6	38.4	33.6

Sources: Angolan authorities and IMF and World Bank estimates.
[a] GDP at market prices
[b] "GNFS" denotes "goods and nonfactor services."
[c] Includes net unrequited transfers excluding official capital grants.
[d] Includes use of IMF resources.
[e] Consolidated central government.
[f] "LCU" denotes "local currency units." An increase in US$/LCU denotes appreciation.

Angola—National Accounts

Part A. Current Price Data (in billions of local currency units)

	Actual			Estimated			Projected	
	1999	2000	2001	2002	2003	2004	2005	2006
Atlas GNP per capita: US$ 1,350 (2005)								
Midyear population: 14.5 million (2005)								
Gross domestic product at market prices	17.2	91.7	197.1	497.6	1041.2	1652.0	2859.7	3538.9
Net indirect taxes	0.6	1.8	3.0	25.2	3.6	−34.9	−6.6	20.1
GDP at factor cost	16.6	89.9	194.1	472.4	1037.6	1687.0	2866.4	3518.7
Agriculture	1.1	5.2	16.1	39.1	86.7	142.5	206.8	322.9
Industry, of which	12.5	66.1	127.9	339.3	701.7	1092.0	2117.0	2552.3
Manufacturing	0.6	2.6	7.6	18.2	40.4	66.1	102.3	131.1
Services	3.6	20.4	53.1	119.2	252.7	417.5	535.9	663.7
Resource balance	−1.1	24.6	3.3	56.4	68.3	263.9	722.2	733.8
Exports (GNFS)[a]	14.8	82.2	151.0	365.9	724.9	1151.2	2102.3	2470.2
Imports (GNFS)	15.9	57.6	147.7	309.5	656.6	887.3	1380.1	1736.4
Total expenditure	18.3	67.1	193.8	441.2	973.0	1388.2	2137.5	2805.0
Consumption expenditures	13.6	53.3	167.4	378.7	841.1	1237.7	1922.0	2307.8
Government	10.3	39.0	68.1	174.3	350.5	501.1	688.5	774.4
Private	3.4	14.3	99.2	204.3	490.6	736.6	1233.5	1533.4
Gross domestic investment	4.7	13.8	26.4	62.5	131.9	150.5	215.5	497.3
Total government investment[b]	2.2	5.6	12.5	33.6	79.0	80.8	134.7	409.5
Total private investment[c]	2.5	8.2	13.9	28.9	52.8	69.7	80.8	87.8
Total fixed investment	4.9	13.8	26.4	62.5	131.9	150.5	215.5	497.3
Total changes in stocks	−0.3	0.0	0.0	0.0	0.0	0.0	0.0	0.0
Domestic savings	3.6	38.4	29.7	119.0	200.1	414.4	937.8	1231.1
+ Net factor income	−3.8	−16.9	−34.4	−71.4	−128.8	−207.2	−355.2	−423.4
+ Net current transfers[d]	0.2	0.3	2.0	1.4	7.4	0.5	2.6	2.2
= National savings	−0.1	21.8	−2.7	48.9	78.7	207.7	585.2	809.9
Gross national product	13.3	74.8	162.7	426.2	912.4	1444.8	2504.6	3115.5
Gross national disposable income	13.5	75.1	164.7	427.6	919.8	1445.4	2507.2	3117.7

[a] "GNFS" denotes "goods and nonfactor services."
[b] Gross domestic fixed capital formation only.
[c] Derived as a residual; includes increase in stocks.
[d] Total net unrequited transfers excluding official capital grants.

Angola—National Accounts (*Continued*)

Part B. Shares of Gross Domestic Product (percentages calculated using current price data)

	Actual			Estimated			Projected	
	1999	2000	2001	2002	2003	2004	2005	2006
Gross domestic product	100.0	100.0	100.0	100.0	100.0	100.0	100.0	100.0
Net indirect taxes	3.2	2.0	1.5	5.1	0.3	−2.1	−0.2	0.6
Agriculture value added	6.3	5.7	8.2	7.9	8.3	8.6	7.2	9.1
Industry value added, of which	72.7	72.1	64.9	68.2	67.4	66.1	74.0	72.1
Manufacturing	3.2	2.9	3.9	3.7	3.9	4.0	3.6	3.7
Services value added	21.0	22.2	27.0	24.0	24.3	25.3	18.7	18.8
Resource balance (X-M)	−6.4	26.8	1.7	11.3	6.6	16.0	25.3	20.7
Exports (GNFS)[a]	86.3	89.6	76.6	73.5	69.6	69.7	73.5	69.8
Imports (GNFS)	92.7	62.8	74.9	62.2	63.1	53.7	48.3	49.1
Total expenditure	106.4	73.2	98.3	88.7	93.4	84.0	74.7	79.3
Government consumption	59.7	42.5	34.6	35.0	33.7	30.3	24.1	21.9
Private consumption	19.5	15.6	50.3	41.1	47.1	44.6	43.1	43.3
Government investment	12.8	6.1	6.4	6.8	7.6	4.9	4.7	11.6
Private investment	14.4	8.9	7.1	5.8	5.1	4.2	2.8	2.5
Gross domestic savings	20.7	41.8	15.1	23.9	19.2	25.1	32.8	34.8
Gross national savings	−0.7	23.8	−1.4	9.8	7.6	12.6	20.5	22.9
Memorandum items								
GDP deflator	888.9	4606.8	9604.4	21178.3	42894.1	61212.3	87851.3	94893.4
Consumer price index	722.4	3070.3	7755.1	16199.7	32130.9	46126.8	56717.9	64049.8
Total GDP (million current US$)	6152.9	9135.1	8936.1	11386.3	13956.3	19799.5	32810.4	44102.7
Per capita gross national product (Atlas method: in US$)	440.0	480.0	520.0	680.0	770.0	960.0	1350.0	1980.0

[a] "GNFS" denotes "goods and nonfactor services."

Angola—National Accounts (*Continued*)

Part C. Constant Price Data (in billions local currency, constant 1997 prices)

	Actual			Estimated			Projected	
	1999	2000	2001	2002	2003	2004	2005	2006
GDP at market prices	1.9	2.0	2.1	2.3	2.4	2.7	3.3	3.7
GDP at factor cost	1.9	2.0	2.1	2.3	2.4	2.7	3.3	3.7
Agriculture	0.2	0.2	0.2	0.2	0.3	0.3	0.4	0.4
Industry, of which	1.2	1.3	1.3	1.5	1.6	1.8	2.2	2.6
Manufacturing	0.1	0.1	0.1	0.1	0.1	0.1	0.2	0.2
Services	0.5	0.5	0.5	0.6	0.6	0.6	0.7	0.7
Resource balance	0.1	0.0	−0.3	−0.1	−0.3	−0.2	−0.6	−1.4
Exports (GNFS)[a]	1.4	1.3	1.3	1.6	1.5	1.7	2.2	2.5
Imports (GNFS)	1.2	1.3	1.6	1.6	1.8	2.0	2.8	3.9
Total expenditure	1.8	2.0	2.3	2.4	2.7	2.9	3.9	5.1
Consumption	1.3	1.5	1.8	1.8	2.0	2.2	2.8	2.5
Government	1.9	1.6	1.2	1.5	1.7	1.8	2.2	2.5
Private	−0.6	−0.1	0.6	0.3	0.3	0.3	0.6	0.1
Gross domestic investment	0.5	0.5	0.5	0.6	0.7	0.8	1.0	2.6
Total government investment	0.2	0.2	0.2	0.3	0.4	0.4	0.7	2.1
Total private investment	0.3	0.3	0.3	0.3	0.3	0.4	0.4	0.5
Total fixed investment	0.5	0.5	0.5	0.6	0.7	0.8	1.0	2.6
Total changes in stocks	0.0	0.0	0.0	0.0	0.0	0.0	0.0	0.0
Terms-of-trade (TT) effect	−0.2	0.6	0.3	0.4	0.5	0.8	2.1	3.0
Gross domestic income	1.7	2.6	2.4	2.7	2.9	3.5	5.3	6.7
Domestic saving (TT adjusted)	0.4	1.1	0.5	0.9	0.9	1.3	2.5	4.2
Net factor income	−0.3	−0.4	−0.4	−0.4	−0.4	−0.5	−0.8	−1.1
GNP at market prices	1.6	1.6	1.7	2.0	2.1	2.2	2.4	2.7

[a] "GNFS" denotes "goods and nonfactor services."

Angola—National Accounts (*Continued*)

Part D. Annual Growth Rate (calculated from data in constant 1997 prices)

	Actual			Estimated			Projected	
	1999	2000	2001	2002	2003	2004	2005	2006
GDP at market prices	3.2	3.0	3.1	14.5	3.3	11.2	20.6	14.6
Agriculture	1.3	9.3	18.0	12.1	12.1	14.1	17.0	14.5
Industry, of which	6.5	3.5	4.1	14.5	3.8	10.6	23.4	17.4
Manufacturing	7.2	8.9	9.8	10.1	12.0	13.5	24.9	13.5
Services	−3.3	−0.3	−4.6	15.4	−1.7	11.3	14.3	5.8
Exports (GNFS)[a]	3.4	−3.7	−1.2	20.7	−3.1	14.4	27.6	12.9
Imports (GNFS)	11.1	6.0	19.1	3.1	10.1	9.4	43.7	37.7
Total expenditure	8.4	10.1	16.5	3.4	12.0	8.2	31.9	32.3
Consumption	4.8	12.7	21.0	−1.0	8.2	10.5	30.1	−11.3
Investment	19.8	2.7	2.7	19.5	23.5	2.1	37.1	150.6
Gross domestic income	21.2	48.0	−7.7	14.4	6.6	21.3	52.0	26.5
Gross domestic saving	−0.2	−19.2	−53.9	144.5	−13.3	14.1	−18.5	184.1

[a] "GNFS" denotes "goods and nonfactor services."

Part E. Period Average Indicators

	Estimate	Projection
	1995–2000	2000–2005
Marginal national saving rate	77.7%	44.8%
Incremental capital-output ratio	4.1	2.5
Import elasticity	0.4	1.4

Sources: Angolan authorities and IMF and World Bank estimates.

Angola—Exports and Imports

	Actual			Estimated			Projected	
	1999	2000	2001	2002	2003	2004	2005	2006
A. Value in current prices (US$ millions)								
Total merchandise exports (FOB)	5157	7920	6645	8166	9515	13474	23724	30335
Principal primary products	5048	7727	6510	8035	9337	13261	23441	29988
Crude oil	4406	6951	5801	7386	8533	12441	22308	28501
Diamonds	629	738	689	638	788	790	1089	1431
Coffee	4	1	0	0	1	0	0	1
Gas	9	37	20	10	16	30	43	55
Refined petroleum products	75	132	93	95	139	148	202	258
Other goods	33	61	42	36	39	65	81	90
Total merchandise imports (CIF)	3109	3040	3179	3760	5480	5832	8667	12514
Memorandum items								
Export volume growth rate	1.4	2.0	−0.9	21.3	−2.7	14.1	28.5	13.2
Import volume growth rate	55.3	4.8	7.9	15.4	29.2	−3.7	43.5	45.5
B. Price Indices (YR97 = 100)								
Merchandise export	99.0	149.2	126.2	127.9	153.2	190.0	260.3	294.2
Merchandise import	92.7	86.5	83.8	86.0	97.0	107.1	111.0	110.2
Merchandise terms of trade	106.8	172.4	150.6	148.8	157.9	177.4	234.6	267.1

Sources: Angolan authorities and IMF and World Bank estimates.

Angola—Balance of Payments (US$ millions at current prices)

	Actual			Estimated			Projected	
	1999	2000	2001	2002	2003	2004	2005	2006
Total exports of GNFS[a]	5310	8187	6848	8373	9716	13797	24120	30785
Merchandise (FOB)	5157	7920	6645	8166	9515	13474	23724	30335
Nonfactor services	153	267	203	207	201	323	397	450
Total Imports of GNFS	5704	5739	6697	7082	8801	10635	15834	21640
Merchandise (FOB)	3109	3040	3179	3760	5480	5832	8667	12514
Nonfactor services	2595	2699	3518	3322	3321	4803	7167	9126
Resource balance	−394	2448	150	1291	915	3162	8286	9145
Net factor income	−1372	−1681	−1561	−1635	−1726	−2484	−4075	−5277
Factor receipts	41	46	49	18	12	33	41	45
Factor payments	1413	1726	1610	1653	1739	2517	4116	5322
Interest (scheduled)	569	597	539	354	243	399	459	688
Total interest paid	569	597	539	354	243	399	459	688

(Continued)

Angola—Balance of Payments (US$ millions at current prices) (*Continued*)

	Actual			Estimated			Projected	
	1999	2000	2001	2002	2003	2004	2005	2006
Net adjustments to scheduled interest	0	0	0	0	0	0	0	0
Other factor payments	844	1130	1071	1299	1495	2118	3657	4634
Net private current transfers	110	126	137	−110	−87	−118	−120	−122
Current receipts, of which	193	186	198	0	0	0	0	0
Workers' remittances	193	186	198	0	0	0	0	0
Current payments	83	60	61	110	87	118	120	122
Net official current transfers	−54	−99	−46	142	186	124	150	150
Current account balance	−1710	795	−1320	−312	−713	685	4242	3896
Official capital grants[b]	7	18	4	10	0	11	6	0
Private investment (net)	2472	879	2146	1643	1652	1414	1639	2237
Direct foreign investment	2472	879	2146	1643	1652	1414	1639	2237
Portfolio investments	0	0	0	0	0	0	0	0
Net Official LT borrowing	−291	−766	−618	−162	178	906	1408	2397
Disbursements (incl. Gap)	1501	1610	1619	1279	1539	2764	3484	3211
Repayments (scheduled)	1791	2376	2237	1441	1359	1857	2076	792
Adjustments to scheduled debt service (Exceptional financing)	791	336	334	334	330	187	335	201
Other capital flows incl. errors and omissions	−739	−630	−985	−1742	−1005	−1970	−5512	−3619
Change in net international reserves (incl. Net IMF) (− indicates increase in assets)	−530	−631	440	229	−443	−1233	−2117	−5112
Memorandum items								
Total gross reserves	496	1198	766	399	800	2034	4147	9261
Total gross reserves (in months' imports G&S[c])	0.8	1.9	1.1	0.5	0.9	1.9	2.5	4.1
Exchange rates								
Annual average (LCU/US$)[d]	2.8	10.0	22.1	43.7	74.6	83.4	87.2	—
At end year (LCU/US$)	5.6	16.8	31.9	58.7	79.1	85.6	80.8	—
Current Account Balance as % GDP	−27.8	8.7	−14.8	−2.7	−5.1	3.5	12.9	8.8

Sources: Angolan authorities and IMF and World Bank estimates.

[a] Goods and non-factor services.
[b] Refers to oil bonus.
[c] "G & S" denotes "goods and services."
[d] "LCU" denotes "local currency units."

Angola—Public Finance (at current prices and exchange rates)

	Actual			Estimated			Projected	
	1999	2000	2001	2002	2003	2004	2005	2006
Government budget (billions LCUs)[a]								
Total current revenues and grants	8.0	46.0	88.9	190.8	394.9	609.7	1085.8	1343.0
Direct taxes	7.4	43.0	78.0	163.6	328.9	522.8	941.2	1160.3
Indirect taxes	0.6	2.8	10.2	25.2	53.1	73.8	109.1	138.4
On domestic goods and services	0.3	1.6	5.8	14.6	30.1	41.9	61.9	78.5
On international trade	0.2	1.2	4.4	10.6	23.0	31.9	47.2	59.9
Nontax receipts	0.0	0.2	0.7	2.0	12.9	13.1	35.5	44.3
Total Expenditures	14.0	53.7	96.0	235.2	461.5	635.7	891.8	1264.1
Total Current Expenditures	8.7	39.9	70.1	174.2	378.1	505.6	720.1	828.5
Interest on external debt	1.6	5.0	9.8	15.5	18.1	31.1	34.8	38.4
Interest on domestic debt	0.0	0.0	0.0	0.0	0.9	6.7	13.8	15.7
Transfers to private sector	0.0	0.0	0.0	0.0	0.0	0.0	0.0	0.0
Transfers to other NFPS[b]	1.8	4.9	3.6	12.5	16.8	32.4	64.0	75.5
Subsidies	0.0	1.0	7.2	0.0	49.5	108.7	115.7	118.3
Consumption	5.3	28.9	49.6	146.2	292.8	326.6	491.8	580.6
Wages and salaries	0.7	5.2	14.9	51.3	125.0	167.0	238.0	314.0
Other consumption	4.6	23.7	34.7	94.9	167.8	159.6	253.8	266.6
Budgetary Savings	−0.7	6.2	18.7	16.6	16.8	104.1	365.8	514.5
Capital Revenues	0.0	0.0	0.0	0.0	0.0	0.0	0.0	0.0
Total Capital Expenditures and other	5.3	13.8	25.8	61.0	83.4	130.1	171.7	435.7
Budgetary fixed investment	2.2	5.6	12.5	33.6	79.0	80.8	134.7	409.5
Other incl. unclassified expenditures	3.2	8.2	13.3	27.4	4.3	49.3	37.0	26.2
Overall balance (− = deficit accrual basis)	−6.1	−7.7	−7.1	−44.4	−66.6	−26.0	194.1	78.9
Changes in arrears net)	1.8	23.3	−2.3	36.8	8.0	−35.0	−23.5	−22.3
Overall balance (− = deficit cash basis)	−4.2	15.7	−9.4	−7.6	−58.6	−61.0	170.6	56.6
Sources of financing (+)	4.2	−15.7	9.4	7.6	58.6	61.0	−170.6	−56.6
Net external borrowing	0.3	−4.3	−10.3	−24.2	36.2	89.6	−91.1	216.7
Net domestic financing	1.3	−11.4	15.4	18.1	22.4	−46.0	−79.4	−433.8
Other	2.6	0.0	4.3	13.6	0.0	17.4	0.0	160.5
Shares of GDP (%)								
Current revenues	46.4	50.2	45.1	38.3	37.9	36.9	38.0	38.0
Current expenditures	50.6	43.5	35.6	35.0	36.3	30.6	25.2	23.4
Budgetary savings	−4.2	6.7	9.5	3.3	1.6	6.3	12.8	14.5

(Continued)

Angola—Public Finance (at current prices and exchange rates) (*Continued*)

	Actual			Estimated			Projected	
	1999	2000	2001	2002	2003	2004	2005	2006
Capital revenues	0.0	0.0	0.0	0.0	0.0	0.0	0.0	0.0
Total Capital Expenditures and other	31.1	15.1	13.1	12.3	8.0	7.9	6.0	12.3
Overall Balance accrual basis (− = deficit)	−35.4	−8.4	−3.6	−8.9	−6.4	−1.6	6.8	2.2

Sources: Angolan authorities and IMF and World Bank estimates.
[a] "LCU" denotes "local currency unit."
[b] "NFPS" denotes "nonfinancial public sector."

Angola—Monetary Survey (in billions of local currency units)

	Actual			Estimated			Projected	
	1999	2000	2001	2002	2003	2004	2005	2006
A. Annual Flows:								
Net foreign assets	5.6	24.7	18.0	46.0	60.1	123.8	185.3	625.1
Net international reserves	1.5	6.3	−9.7	−10.0	33.0	102.9	184.5	410.2
Other net foreign assets	4.1	18.4	27.7	56.0	27.1	20.9	0.8	214.9
Domestic credit	0.8	−14.8	12.3	28.0	38.2	10.0	−14.3	−488.0
To government	0.4	−16.4	6.3	11.5	4.9	−27.2	−68.9	−534.4
To rest of the economy	0.4	1.7	6.0	16.4	33.3	37.2	54.6	46.3
Total assets = liabilities	6.4	10.0	30.3	74.0	98.4	133.8	171.0	137.1
Money and quasimoney	3.4	11.9	25.5	65.6	70.9	66.7	146.6	169.2
Net other liabilities	3.0	−1.9	4.8	8.4	27.4	67.1	24.3	−32.1
B. End of Year Stocks:								
Net foreign assets	5.8	30.5	48.5	94.5	154.7	278.4	463.7	1088.8
Net international reserves	2.8	20.1	24.5	23.4	63.3	174.1	335.0	751.2
Other net foreign assets	3.0	10.4	24.0	71.1	91.4	104.4	128.8	337.6
Domestic credit	1.2	−13.5	−1.2	26.8	65.0	75.0	60.7	−427.3
To government (NFPS)[a]	0.7	−15.7	−9.4	2.1	7.0	−20.2	−89.0	−623.4
Government budget	—	—	—	—	—	—	—	—
Other NFPS	—	—	—	—	—	—	—	—
To rest of the economy	0.5	2.2	8.2	24.7	58.0	95.2	149.7	196.1
Private sector	0.5	2.2	8.2	24.7	58.0	95.2	149.7	196.1
Other financial institutions	0.0	0.0	0.0	0.0	0.0	0.0	0.0	0.0
Total assets = liabilities	7.0	17.0	47.3	121.3	219.7	353.4	524.4	661.5
Money and quasimoney	3.9	15.8	41.4	107.0	177.9	244.6	391.2	560.4
Net other liabilities	3.1	1.2	5.9	14.3	41.8	108.8	133.2	101.1

(*Continued*)

Angola—Monetary Survey (in billions of local currency units) (*Continued*)

	Actual			Estimated			Projected	
	1999	2000	2001	2002	2003	2004	2005	2006
C. Factors accounting for monetary expansion (as % MQM[b])								
Net foreign assets	147.0	192.6	117.3	88.4	86.9	113.8	118.5	194.3
Credit to government (NFPS)	17.9	−99.2	−22.8	2.0	3.9	−8.2	−22.8	−111.2
Credit to rest of the economy	13.6	13.8	19.9	23.1	32.6	38.9	38.3	35.0
Net other liabilities (−)	78.6	7.3	14.3	13.4	23.5	44.5	34.0	18.0
D. Money, credit and prices								
M2/GDP	22.8	17.3	21.0	21.5	17.1	14.8	13.7	15.8
Annual growth rate MQM	680.9	303.7	161.2	158.5	66.3	37.5	60.0	43.2
Annual growth rate private credit	477.4	308.8	275.9	199.9	135.0	64.1	57.4	31.0

Sources: Angolan authorities and IMF and World Bank estimates.
[a] "NFPS" denotes "nonfinancial public sector."
[b] "MQM" denotes "money and quasi money."

References

Adauta de Sousa, M., T. Addison, B. Ekman, and A. Stenman. 2003. "From Humanitarian Assistance to Poverty Reduction in Angola." In T. Addison, ed., *From Conflict to Recovery in Africa*. Oxford: Oxford University Press.

Agénor and Montiel. 1999. *Development Macroeconomics*. Second Edition. Princeton: Princeton University Press.

Aguilar, R. 2001. "Angola's Incomplete Transition." UNU/WIDER Discussion Paper 2001/47, UNU/WIDER, Helsinki.

Alves da Rocha, M.J. 2001. *Os Limites do Crescimento Económico em Angola: As Fronteiras entre o Possível e o Desejável*. Luanda: LAC/Executive Center.

American University. 2002. "A Poverty Profile for Angola." Unpublished manuscript.

Araújo, J.T., and M.N. Costa. 1997. "Patterns of Growth and Structural Change in the Angolan Economy." The World Bank, Washington, D.C.

Auty, R.M. 2005. "The Political Economy of Redirecting Natural Resource Rent to Boost Pro-poor Growth in Angola." Background paper prepared for the Country Economic Memorandum, The World Bank, Washington, D.C.

Barnett, S., and R. Ossowski. 2003. "Operational Aspects of Fiscal Policy in Oil-Producing Countries." In Daves et al., *Fiscal Policy Formulation and Implementation in Oil-Producing Countries*. Washington, D.C.: International Monetary Fund.

Barro, R. 1997. *Determinants of Economic Growth: A Cross-Country Empirical Study*. Cambridge, Mass.: MIT Press.

Bellver, Ana, and Daniel Kaufmann. 2005. "Transparenting Transparency: Initial Empirics and Policy Applications." Forthcoming World Bank Policy Research Working Paper, The World Bank, Washington, D.C.

Bloom, D.E., and J.D. Sachs. 1998. "Geography, Demography, and Economic Growth in Africa." *Brookings Papers on Economic Activity* 1998(2):207–73.

Buffie, Edward, Christopher Adam, Stephen O'Connell, and Catherine Patillo. 2004. "Exchange Rate Policy and the Management of Official and Private Capital Flows in Africa." *IMF Staff Papers* 51.

Busby, G., J. Isham, L. Pritchett, and M. Woolcock. 2002. "Natural Resource and Conflict: What Can We Do?" In I. Bannon and P. Collier, eds., *Natural Resources and Violent Conflict*. Washington, D.C.: The World Bank.

Cain, Allan. 2004. "Chapter 5: Livelihoods and the Informal Sector in Post-War Angola." *Supporting Sustainable Livelihoods*, Monograph 102.

Calvo, Guillermo A., and Carlos A. Vegh. 1999. "Inflation Stabilization and BOP Crises in Developing Countries." NBER Working Papers 6925, National Bureau of Economic Research.

Carneiro, F. 2005. "The Oil Cycle, the Resource Curse, and the Tax-Spend Hypothesis: A VAR Analysis for Angola." Background paper prepared for the Country Economic Memorandum, The World Bank, Washington, D.C.

CEIC. 2004. *Angola Economic Report*. Center of Studies and Scientific Investigation. Luanda: Catholic University of Angola.

Collier, P., and A. Hoeffler. 2005. "Resource Rents, Governance, and Conflict." *Journal of Conflict Resolution* 9:625–33.

Davis, J.M., R. Ossowski, and A. Fidelino. 2003. *Fiscal Policy Formulation and Implementation in Oil-Producing Countries*. Washington, D.C.: International Monetary Fund.

Development Workshop. 2005a. "Angola: Post-Conflict Challenges." Background paper for the Country Social Analysis. Luanda.

———. 2005b. *Terra—Urban Land Reform in Post-War Angola: Research, Advocacy and Policy Development*. Development Workshop and the Centre for Environment and Human Settlements (UK), Development Workshop Occasional Paper No. 5. Luanda.

Dietrich, C., and J. Cilliers. 2000. *Angola's War Economy*. Pretoria, South Africa: The Institute for Security Studies.

Dunovan, C. 1996. "Effects of Monetized Food Aid on Local Maize Prices in Mozambique." Doctoral dissertation, Michigan State University.

FAO. 2003. "Review of Agriculture Sector and Food Security Strategy and Investment."

FAO/MINADER. 1996. *Angola—Agricultural Recovery and Development Options Review (ARDOR)*. Report No. 96/116 TCP-ANG, 5 volumes.

FAO/WB. 2004. "Cooperative Work Program—Economic Sector Work Studies." Luanda.

FAS (Fundo de Apoio Social). 2004. "Vulnerability, Poverty and Social Exclusion." In FAO, *Review of Agricultural Sector and Food Security Strategy and Investment Priority Setting*. Rome.

Fosu, A., and S. O'Connell. 2005. "Explaining African Economic Growth: The Role of Anti-Growth Syndromes." Paper presented at the Annual World Bank Conference on Development, Dakar, Senegal.

Gabre-Madhin and Haggblade. 2004. "Successes in African Agriculture: Results of an Expert Survey." *World Development* 32:745–66.

Gasha, J.G., and G. Pastor. 2004. "Angola's Fragile Stabilization." Working Paper 04/83, International Monetary Fund, Washington, D.C.

Goreux, L. 2001. "Conflict Diamonds." Africa Region Working Paper Series No. 13, Washington, D.C.: The World Bank.

Grion, E.M. 2004. "Mercados e Governos—O Sector Privado em Angola." In CEIC *Relatório Económico de Angola 2003*. Universidade Católica de Angola.

Hausmann, R., A. Powell, and R. Rigobon. 1993. "An Optimal Spending Rule Facing Oil Income Uncertainty (Venezuela)." In E. Engel and P. Meller, eds., *External Shocks and Stabilization Mechanisms*. Washington, D.C.: IADB and Johns Hopkins University Press.

Hausmann, R., and R. Rigobón. 2003. "An Alternative Interpretation of the Resource Curse: Theory and Policy Implications." In J.M. Davis, R. Ossowski, and A. Fidelino, eds., *Fiscal Policy Formulation and Implementation in Oil-Producing Countries*. Washington, D.C.: International Monetary Fund.

Hausmann, R., D. Rodrik, and A. Velasco. 2005. "Growth Diagnostics." Paper presented at the 2005 PREM Week, The World Bank, Washington.

IDR. 2001. "Inquerito aos Agregados Familiares sobre Despesas e Receitas." Instituto Nacional de Estatísticas, Luanda, Angola.

Isham, J., L. Pritchett, M. Woolcock, and G. Busby. 2004. "The Varieties of Resource Experience: How Natural Resource Export Structures Affect the Political Economy of Economic Growth." Middlebury College Economics Discussion Paper No. 03–08R, Middlebury College, Vermont.

Jenkins, P., P. Robson, and A. Cain. 2002. "City Profile: Luanda." *Cities* 19(2).

Katz, M., U. Bartsch, H. Malothra, and M. Cuc. 2004. *Lifting the Oil Curse: Improving Petroleum Revenue Management in Sub-Saharan Africa*. Washington, D.C.: International Monetary Fund.

Kaufmann, D. 2005. "10 Myths About Governance and Corruption." *Finance and Development*. IMF, Washington, D.C.

Kaufmann, Daniel, Aart Kraay, and Massimo Mastruzzi. 2005. "Governance Matters IV: Governance Indicators for 1996–2004." World Bank Policy Research Working Paper 3237, Washington, D.C.

Kyle, S. 1997. "Development of Angola's Agricultural Sector." *Agroalimentaria* 4:89–104.

———. 2002. "The Political Economy of Long-run Growth in Angola." Working Paper 2002–07, Cornell University.

———. 2004. "The Angolan Macroeconomy, Sector Policy and the Agricultural Sector." Unpublished manuscript.

———. 2005. "Oil Revenue and Long Run Growth in Angola." Background paper prepared for the Country Economic Memorandum, The World Bank, Washington, D.C.

Leite, C., and M. Weidmann. 1999. "Does Mother Nature Corrupt? Natural Resources, Corruption and Economic Growth." IMF Working Paper WP/99/85.

Loayza, N., A.M. Oviedo, and L. Servén. 2005. "The Impact of Regulation on Growth and Informality: Cross-country Evidence." Policy Research Working Paper No. 3623, The World Bank, Washington, D.C.

Malaquias, A. 2000. "Ethnicity and Conflict in Angola: Prospects for Reconciliation." In C. Dietrich and J. Cilliers, eds., *Angola's War Economy*. Pretoria, South Africa: The Institute for Security Studies.

Mauro, P. 1995. "Corruption and Growth." *Quarterly Journal of Economics* 90:681–712.

McMahon. 1997. *The Natural Resource Curse: Myth or Reality?* Washington, D.C.: The World Bank Institute.

Mellor, J.W. 1995. *Agriculture on the Road to Industrialization*. Baltimore, Md.: Johns Hopkins University Press.

MINADER. 2003. *Linhas Gerais de Orientacao do Programa Global do Sector Agrario para o ano agricola 2004/2005*. Luanda, Angola.

———. 2004. *Programas Multisectoriais em Curso 2003/04; Projectos em Execucao no Ambito do Programa Economico do GOA 2003/04*. Luanda, Angola.

OECD. 2005. *African Economic Outlook*. Paris.

Pacheco, F., 2005. "Inventário dos programas e projectos em curso no sector rural em Angola." Unpublished manuscript.

Pearce, J. 2004. *War, Peace and Diamonds in Angola*. Pretoria, South Africa: The Institute for Security Studies.

Rivera-Batiz, F.L. 2003. "Democracy, Governance, and Economic Growth: Theory and Evidence." Deparment of Economics, Columbia University, Processed.

Robson, P. 2005. "Angola: Vulnerability, Resources and Conflict." Background paper for the Country Social Analysis, Development Workshop, Luanda.
Rodrick, D. 1999. "Institutions for High Quality Growth: What They Are and How to Acquire Them." Harvard University, Processed.
Rodrick, D., A. Subramanian, and F. Trebbi. 2002. "Institutions Rule: The Primacy of Institutions over Geography and Integration in Economic Development." Harvard University, Processed.
Sala-i-Martin, X., and A. Subramanian. 2003. "Addressing the Natural Resource Curse: An Illustration from Nigeria." Unpublished manuscript.
Sarraf and Jiwanji. 2001. "Beating the Resource Curse: The Case of Botswana." Environmental Economics Series, Paper No. 83, The World Bank Institute, Washington, D.C.
Shen, J.G. 2002. "Democracy and Growth: An Alternative Empirical Approach." BOFIT Discussion Papers, No 13, Bank of Finland, Institute for Economies in Transition.
Tijerina-Guajardo, J.A., and J.A. Pagán. 2003. "Government Spending, Taxation, and Oil Revenues in Mexico." *Review of Development Economics* 7:151–64.
UNAIDS. 2004. *Angola: UNAIDS/WHO Epidemiological Fact Sheets on HIV/AIDS and Sexually Transmitted Diseases—2004 Update*. Geneva: UNAIDS/WHO.
UNCTAD. 2004. *World Investment Report 2004: The Shift Towards Services*. New York and Geneva.
UNDP (United Nations Development Programme). 2005. "Defusing the Remnants of War Economics, Economic Report on Angola in 2002–2004. Draft, UNDP, Luanda.
UNICEF and INE (Instituto Nacional de Estatística). 2003. *MICS—Inquérito de indicadores múltiplos*. Luanda.
Van Wijnbergen, S. 1996. *Aid, Export Promotion and the Real Exchange Rate: An African Dilemma?* World Bank Discussion Paper No. 199. Washington, D.C.: The World Bank.
World Bank. 1990. *Angola: An Introductory Review*. Washington, D.C.
———. 1994. "Angola: Strategic Orientation for Agricultural Development, An Agenda for Discussion." Washington, D.C.
———. 2003. *Transitional Support Strategy for the Republic of Angola*. Report No. 25471-ANG. Washington, D.C.
———. 2005a. *Angola: Public Expenditure Management and Financial Accountability Review*. World Bank Report 29036 AN. Washington, D.C.
———. 2005b. "Angola: Towards a Strategy for Agricultural Development, Issues and Options." Unpublished manuscript.
———. 2005c. *Country Framework Report—Private Solutions for Infrastructure in Angola*. Washington, D.C.
———. 2005d. "Country Social Analysis." Background paper prepared for the Country Economic Memorandum, Washington, D.C.
———. 2005e. *Equity and Development, World Development Report 2006*. Washington, D.C.
———. 2005f. "Managing Angola's Oil Wealth." Background paper prepared for the Country Economic Memorandum, Washinton, D.C.

Eco-Audit

Environmental Benefits Statement

The World Bank is committed to preserving Endangered Forests and natural resources. We print World Bank Working Papers and Country Studies on 100 percent postconsumer recycled paper, processed chlorine free. The World Bank has formally agreed to follow the recommended standards for paper usage set by Green Press Initiative—a nonprofit program supporting publishers in using fiber that is not sourced from Endangered Forests. For more information, visit www.greenpressinitiative.org.

In 2006, the printing of these books on recycled paper saved the following:

Trees*	Solid Waste	Water	Net Greenhouse Gases	Total Energy
203	9,544	73,944	17,498	141 mil.
*40' in height and 6–8" in diameter	Pounds	Gallons	Pounds CO_2 Equivalent	BTUs

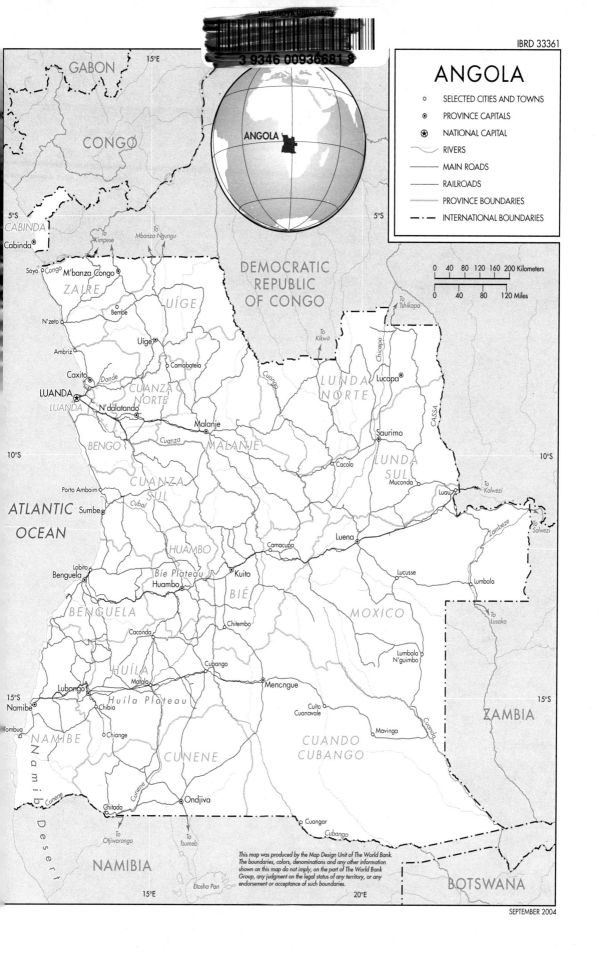